THE EMPIRICAL GOD

by: Casey Tucker

© Casey Tucker, 2023

"A book must be the axe for the frozen sea within us."
Franz Kafka

For my younger brother, Luke Tucker. May you rest in peace and remain in our hearts forever.

Table of Contents

INTRODUCTION	5
THE FIRST STEP TOWARD PROPER ORIENTATION	17
THE FOUNDATION	25
VALID QUALITATIVE METRICAL ORIENTATION	36
REBUILDING FROM THE COGITO	45
UNDERSTANDING LANGUAGE AS A BARRIER	53
SENSING, KNOWING, AND TIME	74
EVOLUTION AND CONTEXT	86
EVOLUTION OF CONSCIOUSNESS	99
DEFINING GOD	114
REFERENCES	125
ACKNOWLEDGMENTS	126
ABOUT THE AUTHOR	127

INTRODUCTION

The word "God" is an allegorical representation of the most essential qualities at the edges of human understanding. These qualities include the first cause of all existence, the end of all existence, the moral authority, and the determinant of human fortunes and fate. But above all, "God" has been a confounding existential puzzle for inquiring minds that seek more profound meaning in this reality. Any given depiction of who/what God is must mirror the grandiose qualities this word embodies and be accompanied by a wealth of substantiation. A spectrum of myths and stories are insufficient justifications for substantiating an all-encompassing scope of time and guiding principles. Thus, we must be able to formulate an encyclopedic reasoning that demonstrates the grandiose qualities of "God." In other words, we must turn toward a comprehensive, empirically derived God.

If we wish to discover the answers to life's most puzzling existential questions, like: "Who is God?" Then we must abandon our generation's cultural habits of reducing and simplifying. The construction of meaning must adhere to the interminably dissipating nature of original forms found in this universe. Regardless of how confounding original forms of matter may be, we must apply unrelenting intellectual tenacity to form deeper meaning. Meaningful answers to life are not discovered through conforming to standardized cultural habits of pursuing simplicity. All forms of life pose insurmountable complication and wonder. How is it still observed that people contemptibly reduce themselves to arbitrary forms? As exemplified by repeated idioms like "a cog in a wheel."

Human beings can be the pride of Mother Nature; we are the most sophisticated manifestation of creation. We are nature's realization of the creature who realizes itself; we have the potential to be the modern triumph of restored dignity to nature. The sooner we accept this responsibility, the sooner we can recognize the conscious potential of our existence.

Habits of simplicity lead to cerebral pathogens like the philosophical doctrine of nihilism, which functionally harms the cognitive development of the self and the collective. Nihilism is "extreme skepticism maintaining that nothing in the world has a real existence, and thus, no meaning" (Oxford Languages). Accepting nihilism as truth is the most abhorrent cerebral crime of human existence; such reductions of the self are naive, lazy, and untrue. Assigning value to the mind entails the absolute consideration of itself amid all the complexities of its aggregate in which it is integrated. Consider, for a moment, the commonly taught reduction of a given mathematical fraction. The practice of mathematical simplification poses a particular temptation to accept the simplified form as superior. In actuality, the utmost reduced fraction bears no more correct value than the multitude of acceptable, non-reduced forms of a given mathematical fraction derived from an equation. If we are to follow the proper procedure of discovering the answer to a mathematical equation that results in a fraction, reducing it further would deviate it from the value that provides semblant context to its origin. Thus, qualitatively, it becomes *generalized* rather than *refined*. This cultural standardization of simplification has become a growing issue in quantitative and qualitative thought processes. This quantitative practice of fraction reduction is procedurally congruent with the qualitative practice resulting in nihilism.

    The notion that it's acceptable to invent a body of meaning which asserts there is no meaning is not only inherently ironic but a symptom of the underlying complication causing poor human understanding. A symptom of poor human understanding is often identifiable in an idea's lack of nuance. Lack of nuance is where philosophy and science offer great value to human understanding regarding more cognizant orientation with reality. In the conceptual space stimulated by philosophical thought, ultimate ideas such as "God" are developed in a more nuanced form. But perhaps the unfortunate crime of philosophy thus far is its incommunicability with those who haven't been trained in the discipline. You might read original work or commentary on familiar names such as Nietzsche, Hegel, Descartes, or Kant and feel an overwhelming sense of uncertainty. Or even worse, you may have done the arduous work to understand the content yet find yourself less sure of reality than before. Incommunicability is prevalent within philosophy primarily because Western philosophy focuses on answering essential yet unanswerable questions. Examples of such questions would be: "How do we know what we know?" "How do we live a moral life?" "How do we organize reasoning?" On the positive side, unrestrained investigation of such questions has provided much utility in producing an ambitious scope toward pursuing knowledge.

As a result of philosophical ambition, vital thinkers have advanced the historical progression of Western ideation. However, the popularly discussed and accepted doctrines need to provide more functional guidance. Human understanding currently features an increase in depictions of reality devoid of proper contextual orientation with its own logical, ideological, and sensuous environment. More specifically, human understanding lacks critical nuance. Philosophy has a distinct duty to provide practical guiding wisdom toward the progressing collective of human understanding. Over the progression of Western thought, philosophical wisdom has struggled to maintain a critical focus consistent with and relevant to our evolving knowledge of the natural world and technological capacity. Human comprehension of the environment often lacks critical orientation with reality in Western doctrines. This lack has not improved with our current bodies of wisdom. The essential goal in writing this book is to re-orient human understanding with the growing parameters of our holistic environmental knowledge, which provides the essential context of our existence. In re-orienting ourselves with our environment, the chaotic nature of differentiating grand notions, like the notion of God, begins to materialize.

In this book, the steps taken to understand what the human environment entails goes as follows: A reconsideration of "truth" and objectivity, a conceptual common ground for every human being, a qualitative metric system, an investigation of language and its impact on conceptualizations of reality, an advancement of a more fundamental understanding of nature, and understanding theoretical knowledge in larger temporal frames that reveal errors that have plagued the conceptual development of consequential ideas like "God." Such endeavors are critical in adequately guiding the construction of theories that maintain contextual orientation with the environment. The profundity of truth lies within the accurate expansion of the philosophy of mind with and as its environment. Philosophy must have the orienting questions and answers for our species' most prominent issues.

Additionally, the development of Western thought hasn't effectively prevented existential suffering from staking itself in our culture. Addressing such suffering is another essential goal in writing this book. Providing the mind with the most holistically accurate depiction of its existential context will provide meaning that should curb existential suffering. Philosophy hasn't sufficiently accomplished the task of providing the human mind with a comprehensive depiction of its environment to date. Ultimately, careful synthesis of understanding the environment will formulate a more complete, nuanced, and practical understanding of the human context, human purpose, and God.

The consequence of improper orientation to human reality leads to a chaotic fallout in agreements on truth and drastic misconceptions of fundamental concepts, such as the ideation of God. Lack of contextual orientation with existence in a holistic sense leaves an observably chaotic state in human meaning and existential being. A holistic understanding of reality would begin with evaluating the function of existence in the largest possible frame. The largest possible frame entails evaluating the vast constellation of the universe containing an unlikely speck that actively beholds the entirety of our species. The whole of known meaning, value, and truth exists within the speck's field of time and space. This collective body of comprehension contains an amalgamation of applied delusions disguised as doctrines, ideologies, and morals. Most of these notions mistakenly propose a version of 'truth' representing an absolute form. The chaotic, relative, and randomized fusion of contradicting absolutes enacts a methodically dissipating construct of meaning. Consequently, we are left with a body of human meaning lacking agreement on what "holistic" even means.

When two (or more) sets of interactive minds adopt exclusively different sets of absolute truths, they inflict disagreement and conflict, resulting in the worst possible consequences. Hence, the long history of human war and senseless violence to date. The observable fact that war ever existed demonstrates the importance of comprehensively considering how, why, and what ideas are accepted. Human history has consistently observed opposing ideas or beliefs as the cause of abhorrent tragedies. Ideas, doctrines, and manifestos can cause any human being to abandon their engrained survival instincts to eradicate the experience of anyone who accepts an opposing notion. If humanity is to promote survival, it is undoubtedly preferable to deploy systems of meaning that erode perpetual, stark disagreement between exclusive relative truths, especially considering our growing technological capacity to exterminate all life.

As posited by Descartes in the 17th century, our philosophical questioning must congruently match the advancement of scientific knowledge. In the last century alone, we have acquired context-altering knowledge and technological capacities like social media, DNA sequencing, artificial intelligence, and advanced nuclear-powered weaponry that pose an increased threat and complicates the current relationship between scientific knowledge and collective guiding wisdom. With an unprecedented set of scientific knowledge and technological power driving conceptual chaos, we must advance the applicability of our guiding bodies of wisdom.

When operationalized, accepted truths present purely relative forms of understanding this world, the unfortunate condition of relativity further confuses our understanding of truth itself. The benchmark of truth being relative when operationalized is the metric which renders intelligible bodies of ideas insufficient. **Because human understanding is purely relative, we must make a substantial effort to understand all we are relative to.** The erosion of collective discrepancy in large frames of meaning marks success rather than failure, which promotes accuracy and cohesion. I am not suggesting there only be one body of meaning that we must select and abide by for the sake of uniformity. But meaning should not permit an unlimited array of conflicting notions that take the form of truth, as seen in today's exchanges. We are delusional if we continue to display complacency over the current discrepant truth condition, and there is an increasing urgency to identify and solve these problems collectively.

Recognizing this condition is becoming exceedingly important as the variables of human life and the breadth of human knowledge continue to expand. Still, new postulations are necessary for a system of knowledge that is expanding, and disagreements in systems of meaning are inevitable. However, disagreements should present themselves when discussing historical and contemporary doctrines, not when comparing doctrines with your neighbors. Temporal relativity in truth is a necessary feature of human meaning, given that it must be constructed over time through arduous effort. However, an extensive collection of contemporary doctrines should not contain the current magnitude of discrepancy between the operationalized truths of our current doctrines. May this be one of the many directed purposes of philosophical thought: **to mitigate the unfolding magnitude of disparity within the content recognized as true.**

My report on the status of truth within human meaning does not offer a grand hypothesis of "objective" truth distinct from the human being but rather acknowledges how truth is operationalized. What body of reason even permits such a concept as objectivity? What knowable matter is distinct from the human being? As any notion becomes socially and formally accepted amongst select factions of humanity, it holds consequential value in its capacity to govern the group accepting this contingent truth authoritatively. Thus, an authoritative group accepting a system of meaning renders a doctrine to be true, even though it may not be "correct." Extreme relativity in applied 'truth' may only be a result attributable to the unfortunate circumstances of the human being. A multitude of distinct systems of meaning are inevitable, but we are not impotent in solving the issue of chronic relativity in truth. We are only human, and these problems must be taken on methodically over extensive periods of time.

Human structures of meaning have always presented a constellation of relative conflicting notions of *truth* when operationalized. Truth being relative has always been the functional condition pertaining to all human meaning. The observation of unregulated social media platforms has provided further clarity to this observable condition. The digital information cartels on these platforms prominently display the canvas of an exponential chaotic fallout in truth. On these platforms, the conglomeration of relativity in meaning is displayed with extraordinary accentuation. This digital context provides a more explicit and transparent presentation of the human conglomerate of conception relative to truth.

However, my concern is not to dissect social media's impact through a particular scholarly lens like sociology or psychology. Such platforms display an accentuated representation of the fallout of truth. My concern in this text is comprehensively confronting problematic elements of the human environment. As a consequence of simply existing, humankind is tasked with providing a rationale for why one is not arbitrary. The human need to invent meaning may be the world's greatest mystery. No single human being is to blame for their condition; the existential mystery is the price of admission for the sentient being. The current host of answers to the problem of existence itself is built on real intellectual work and elaborate detail, but it remains insufficient. This becomes quite clear when we observe the mind's tendency to persistently doubt itself. It is the condition of the human being that we are burdened with complete responsibility in discerning ultimate existential truth; it sits as an omnipresent dreadful task to all humans, to have only been masked temporarily by God(s). But God(s) only serves to alleviate this burden, like a pain reliever for the soul, and like pain relievers, these deficient ideas are finite.

Human thought has consistently shown to be exceptionally hopeful for the achievement of discovering *truth*. This text has been written to give back hope in the feasibility of this achievement to those who lost it. A critical benchmark in the history of humanity is humanity's long history of changing its beliefs but never changing its hope. However, it is a pertinent fault of our generation that a massive diffusion of responsibility exists in the pursuit of understanding what *truth* is. The diffusion of responsibility instilled in the West through current dogma leaves the mass of Western civilization unprepared for the coming storm of critical decision-making it is responsible for in the coming years. The most fundamental responsibility of the human is to obtain some semblance of proper orientation to itself and where it stands existentially. A concrete understanding of its place in the environment is the starting point in formulating the most accurate *truth* of our existential being.

Throughout this book, I will affirm, question, and reimagine many ideas methodically developed over the last five centuries of Western philosophical thought. It is important to understand that each idea and chapter entails a fundamental assertion of reality intended to contextualize the human experience holistically. Through this, you may find that faith is not necessary to believe in God. Empirically driven awareness of the human context provides sufficient material to understand who God really is. As you may find, using terms that carry sentiments like "spiritual" and "mystical" consistently indicate concrete components of the human experience that are yet to be fully understood.

THE FIRST STEP TOWARD PROPER ORIENTATION

"There is nothing deep down inside us except what we have put there ourselves."
–Richard Rorty

Understanding the layers between the subject (human being) and object (objective reality) of the human environment is essential for establishing existential contextual orientation. You might be surprised to discover the volume of observable layers between the subject and the object. Functionally, truth depends on the complex dynamics of the individual. If we observe the individual more thoroughly, a holistically orienting lens of meaning becomes more available. The individual (subject) is composed of inherited genes, culture, language, perception, intellect, history, and state of mind. Each constituent part composes the subject and actively impacts the operationalization of conceptual notions of perceived "truth."

An omnipresent element in the environment of human conceptualization is the notion that human meaning is exclusively subjective. This important assertion establishes the baseline environment for human sensing. Observing the idea of science and how it is operationalized helps us understand the notion of omnipresent subjectivity better. Science is a crucial tool for humankind's attempts at orienting with reality and is very important to this book. The intent behind the creation of science was to create a breadth of information we can categorize as objective. Therefore, it is crucial to pry science open to understand why subjectivity and objectivity present a false dichotomy.

The scientific method is a strict procedural guide for the subjective conscious mind. The eight-step scientific process is laid out as follows:

1. Inquiry: Questions about the experience, expressing a yearning to find the best explanations and understanding the external-temporal world.
2. Awareness of the context: It is essential to understand the context in which a question or element is embedded to interpret a given puzzle correctly and adequately.
3. Observational skills: An observer must collect sufficient data on the element in question while considering its context and given parameters.

4. Insight: This is where an educated guess or hypothesis is made by synthesizing the data that has currently been discovered.
5. Reasonableness/Rationality: This is where self-check is done, where we check the adequacy of the investigation and whether the hypothesis is appropriate.
6. Good Judgment: Another self-check to ensure the observations and explanations align with the context and hypothesis.
7. Perseverance and Persistence: Repeating all the steps so far to persistently check all the experiment elements.
8. Responsibility: Deciding what to do about the conclusions reached in step six, which entails all decisions, changes, and results.

Each step requires the intervention of the subject conducting any given research matter. The observer is a story of its own, which contains its personal history, attachments, dispositions, language, neurochemical balance, emotional state, perceptual acuity, and adopted systems of meaning that determine its lens for the conceptualization of truth within this reality. The listed components of the observer can be appropriately reduced to the composite of inherent omnipresent bias. It functionally operates as a lens determining the potential variations in *something* perceived. Indeed, an objective view of judgment would be devoid of such properties, but the human condition renders this impossible. All sentient beings are born into this essential condition. This condition renders objective judgment as nothing more than a fable.

The focal point is that the doctrine of scientific understanding has often led to a standard misconception that objectivity is attainable through this methodology. It is important to distinguish my message apart from the dismissal of science. Science has been a candle in the dark for human civilization and is the foundation for discovery and progress. It functionally serves as a pillar of sense-making in the world. But it is essential to note that it reflects the negative aspects of its creators in one way or another. The misconception of objectivity stands as the magician's trick to be pried out of the trojan horse in the doctrine of science. The general reference to science ought to be seen for what it is. Science is a formatted methodology that serves the function of optimizing knowledge that arises from external sense experience. Accepting absolute trust in discoveries resulting from "objectivity," causes you to lose sight of reasonable skepticism. Losing reasonable skepticism could evolve into blind acceptance caused by excessive faith in the illusion of objectivity. Ultimately, if we accept the notion of objectivity, we increase the odds of deviating from the disciplined nature science intended to uphold.

The propensity of the human being to believe that "objective" or "absolute" truth is an attainable reality is a root causal mechanism that produces problematic forms of cooperation on massive scales. To understand grandiose concepts like "ultimate truth" or "ultimate reality," you must realize that a composition of properties qualifies it as grandiose. Utilization of the scientific method is designed to facilitate intelligible perpetual skepticism; this intelligible skepticism fuels its success, not its assumed objective nature. You must not forget that falsifiability is a necessary feature of any scientific discovery. Ironically, this feature of falsifiability gives science more validity, not some notion that it is impervious to being incorrect. The longest-standing ideas that arose out of science are presumably the hardest to falsify, which suggests the probability of them being true to be almost certain. Unfortunately, *truth* being "almost certain" is the best we can do in our judgments.

In its moral application, the acceptance of objectivity has a similar toxic effect that dogma might have on the conscious individual. Firstly, each portrays itself as an absolutely valid doctrine of meaning that deludes the individual sentient mind into justifying devotion to its supposed principles. Secondly, it has the effect of diffusing essential responsibility of the subjective conscious intellect. Consequently, this can be inherently misleading when we stand in the face of moral decision-making. The deceptive nature of this illusion is the point of concern here; it disorients us from the true nature of what it means to be human. And what it means to be human is to rely entirely on our functionally subjective framework. Whether it be an arbitrary dogmatic system of morality or the devotion to the imaginary construct of objectivity, each presents a subtly misleading protocol to follow within its respective doctrine's guidelines. Misconceptions of objectivity can destructively cause our consideration of the consequential environment to be of secondary concern to the presumed moral protocol of the respective dogma, whether religion or the illusion of objectivity.

No matter how carefully reliant you are on the guidance of the external world, everything starts and ends with the subjective mind, all its functional components, and procedures. The subjective condition is that all meaning originated from a human mind as the sole proprietor of materialized ideation. The subjective condition entails that the mind is the alpha and the omega, given that the conscious observer determines all origination, observation, synthesis, and acceptance. This understanding is at the heart of why human beings are burdened, in the absolute sense, with profound and consequential responsibility. We exist in this landscape where reality determines the quality of our existence, yet we determine this reality.

Empirical application performed by the intelligible skeptic who is aware of the human condition is undoubtedly preferable in facing complex consequential questions. Absolute responsibility lies within the subjective mind; we must realize and accept that the conscious subjective is the sole proprietor of all meaning in this world. Objectivity has yet to be achieved, but the scientific process is the closest thing we have, which suggests that success in originating any meaning must be systematic and guided.

The condition of methodically constructing meaning is analogous to solving a maze from the vantage point within the maze's walls. The analogous nature of the formation of essential meaning and the act of solving a maze lies in the dynamics of the investigative procedure each presents. You can only discover the next progression of a maze by actively taking steps regardless of the direction. When wandering through a maze, the wanderer inevitably runs into some form of a dead end or redirective cue. Progress in finding the solution to a maze depends on the process of elimination through discovering which paths lead to dead ends. You only get to the end of the maze by realizing, through empirical observation, which routes are incorrect, thus narrowing the scope of correctness toward a conclusion of optimal correctness. Discerning incorrectness through principled redirective cues is essential to find the most accurate interpretations of what is. Similar to how a maze contains a solvable end, meaning formed from the mind reaches measurable accuracy when congruent with environmental cues.

The collective of human thought has no such reference point because it has, unfortunately, miscategorized schools of thought and perceptual notions that have tremendous potential to be applied as a helpful reference point. The extreme discrepancy between relativity within humanity's current form of operationalized meaning and scientific knowledge without adequate guiding wisdom gives reason to reorient the mind. In terms of humankind's understanding of its existential context, this discrepancy symbolizes the mark of a dead end or at least a redirective cue. This cue presents a motion for the mind to retract and discover a new directive reference point to reorient the collective mind with its environment.

THE FOUNDATION

"If you would be a real seeker after truth, it is necessary that at least once in your life you doubt, as far as possible, all things."
–René Descartes

So, where does the intelligent skeptic begin building the most accurate depiction of the environment? The best answer may lie in the wisdom provided by René Descartes centuries ago. In *Meditations II,* Descartes' methodology of absolute doubt serves a tremendous functional purpose in formulating a comprehensive starting point for focus and direction of thought. This form of absolute doubt, initially posed by Descartes, referred to as "Cartesian doubt," alludes to a method of doubting all forms of knowledge. The purpose of applying Cartesian doubt was to produce a point of foundational certainty that may serve as a universal reference point for human thought. The benchmark indicating the necessity of this exercise was that the philosophical focus of Descartes' time was obsolete compared to the more advanced expansion of knowledge driven by the sciences.

The current discrepancy between the lack of guiding wisdom and the more advanced nature of scientific knowledge indicates that this exercise should be re-created considering our evolved understanding of the world. Within this space lies a newfound approach with the potential to augment philosophy's focus and expand its scope toward a more productive end.

In the 1600s, the methodology of cartesian doubt was utilized for the paradigmatic establishment of particular doctrines. The history of its application within these territories is especially worn; not only has it been worn, but it has been worn in insufficient applications that have yet to realize the profound potential of this methodology. Western thought has yet to see the functionality of this endeavor applied to the collective human concept of truth. To elaborate on cartesian doubt, imagine taking a profound odyssey leading you into a dimension of existence devoid of any reference point that provides meaningful orientation. Cartesian doubt is the act of doubting the certainty of any past moment leading up to the present, doubting the present leading to the future, doubting that the sun will rise tomorrow, doubting all social conventions, doubting the laws of physics, doubting fundamental arithmetic and logic, doubting the certainty of language, doubting the senses, doubting conceptions of the self, and doubting your breath.

This odyssey of doubt left Descartes suspended in a void where he had doubted until there was nothing left to doubt. Within this space, something became evident; no matter how hard he tried to doubt, the doubter within the mind was always present. The only occupant in this existence is the thinking mind which enacts this task of doubting. Once the doubting mind has dismissed all relative substance, the only thing you cannot doubt, no matter how much you are doubting, is whatever is producing this doubt will remain present so long as it is alive. The fact that you are doubting is evidence of *something* undoubtedly existing. How else would this doubt be possible? It certainly can't be from nothing. This instance of this realization is referred to as the cogito.

May this be the highest form of certain truth: If we are alive in this realm of existence, thought and being are two separate properties of humankind. The cogito moves forward the notion that experiential consciousness will always be a property in the living human being. **Let the properties of thinking and being be the common ground for all of humanity.**

This realization of the cogito is the initial pathway in the maze of meaning-making without a dead end. The cogito is where Descartes coined the famous term "*je pense, donc je suis*," or "I think, therefore I am." The undoubted nature of this foundational point of knowledge ought to be accepted by all doctrines as the fundamental starting point for meaning, wisdom, and all knowable things. The phrase "I think, therefore I am" effectively confirms the certainty of the phenomenon of conscious experience. What is thinking, and what is being, if not the qualities representative of this phenomenon? The best definition we can assign to this phenomenon of consciousness is: "the awareness by the mind of itself and the world" (Oxford Languages). By enacting Cartesian doubt, we effectively performed an experiment that embodies this definition.

The cogito similarly affirms the foundational understanding of "being and thought" to be certain in the eyes of the conscious mind, thus introducing consciousness to itself. The value of this discovery is self-evident, which inherently renders it a firm cornerstone in the body of human knowledge. At the foundation of understanding, consciousness is the essence of the phenomenological. Apart from its certainty, the essence of consciousness entails qualities of sanctity, power, universality, and wonder.

Cartesian doubt's upheaval in the body of human wisdom left an important suggestive wake in Western thought. The method not only reinforces that reconfiguring your entire system of meaning and sense-making is feasible, but it is also necessary to obtain clarity and certainty amid a progressing world. If the foundational building blocks of our systems of meaning are not tested with relevant reference points, there can be no reason to be certain of them. In the cogito, however, there were no reference points aside from the doubting mind; therefore, it establishes the initial point of reference. The cogito confirms the essential condition that any given system of meaning has a formal starting point.

Cartesian doubt may be inherently uncomfortable because of its disorienting nature, but it is essential to confront the nature of the self with the relevant spatio-temporal context. The term "spatio-temporal" is a direct reference to the context that contains the self-experience within perceptively understood relevant dimensions encompassed by space and time. Spatio-temporal, then, identifies the specific parameters of the environment that uphold and govern the rules and experience of the conscious human mind. The composition of the consequential spatio-temporal context is the fundamental infrastructure that determines elements of existence itself; therefore, it must be understood to be properly oriented with reality. Similarly, you must understand the essence of experiential consciousness to be oriented with the reality of the self.

To progress Western thought most accurately, we must understand the mistakes made by Descartes that caused the cogito to be less effective, like postulating rationalism as a reasonable doctrine to conclude from the cogito. Rationalism is "the theory that reason rather than experience is the foundation of certainty in knowledge" (Oxford Languages). In traditional Philosophy, rationalism is contrasted with empiricism, defined as "the theory that all knowledge is derived from sense-experience. Stimulated by the rise of experimental science, it developed in the 17th and 18th centuries" (Oxford Languages). The origination of these antithetical doctrines generated a centuries-long debate over the superior theory of knowledge. The perpetuated distinction between these two theories has effectively delayed our understanding of the human context by stimulating a false dichotomy. Consequently, this false dichotomy stunted our growth in understanding how we come to "know" and how we ought to achieve certainty.

At the inception point of Descartes' construction lies a pure assumption regarding the context of the mind. It was entirely missed that the endeavor of absolute doubting is an instance of functionally utilizing your senses, which isn't congruent with rationalism's tenets at all. A consequential mistake in *Meditations II* is Descartes' excessive distinction between thoughts and senses. Generally, this delineation is chalked up as a "common sense" way to mentally categorize the elements of human perception. Applying this delineation in *Meditations II*, Descartes operates under the notion that any activity inside the brain is solely a thought, and anything outside the brain is exclusively reserved as a sense. According to Descartes, the only differential marker between a thought and a sense is the distinct point of origin where the object of cognition occurred. Descartes' delineation draws an assumed omnipresent universal, which states that an object of sense entails the detection of substance 'outside' the mind.

If we can sense (detect) contents of the external context, how does that not entail that we can't sense (detect) an internal context? The notion of an externally sensible world inherently entails an internally sensible world. Can we no longer reason that an external implies within itself an internal? If there were no contrary form to the term "external senses," it would be an arbitrary and purposeless linguistic point of reference. Of course, using the orienting phrases "external" and "internal" is not arbitrary because they functionally identify accurate representations of the context that encapsulates the human mind. When you consider that there exists an external sense experience, the inherent, deducible counterpart is that it must entail an internal sense experience.

If anything, the cogito verifies that once consciousness begins to realize itself, it does so through a form of internal sense. The experience may have been generated by thinking, but the thinking produced an experience. The fact of the cogito as a sensed event renders it a form of empirical discovery rather than a mystical conjuring of metaphysical material referred to as thought. Any theory of how we know must be concurrent with this notion. The critical error is the assumption that detecting a synthesized product of thinking results exclusively from thought and not a combination of thought and sense. Understanding that our perception entails both an internal experience and an external experience further provides a context for how we should envision this phenomenon.

The cogito is derived from the internalized sense experience initiated by the internal gaze of absolute doubt. The cogito quite literally displays our capacity to utilize the imagination in a way that simulates a sensed experience. Regardless of the experience's origin, it is a sensuously detectable experience. You can imagine the mechanism of the mind enacting this doubt from an 'aerial' view. This projected 'aerial' view puts you at a third-party vantage point to observe the doubting from a birds-eye perspective. It may not be the conventional act of sensing external reality as empirical observations are commonly portrayed. But the mechanism does functionally take the form of an empirical observation of the most fundamental state of the human mind. If you don't experience absolute doubt or observe the experience, you will not come to this realization, as seen in Descartes' cogito. Observing consciousness through Cartesian doubt is the process of observing the mechanism of your existence from an internal vantage point. This entails we need to expand the definitional understanding of what it means for a human being to sensuously detect.

Once the content of reason expands beyond the cogito, the method of this reasoning doesn't shift as it must. The deduction "there is always thinking; therefore, there must be *being*" is self-evident because the reduction of thought is the essential component of the conscious mind. However, claims that follow the cogito do not feature logically self-evident precedents. Once we move beyond the essential components of the conscious mind, we expand beyond the self-evident and the parameters of reference points by volume. As reasoned above, the cogito was an observed experience, and experiences are only realized through a form of sense and need additionally verifiable reference points. Determining the nature of the method used in the cogito requires a more critical synthesis of additional relevant measures. This mistake in Descartes' *Meditations II* led to the ultimate misunderstanding of the empirical nature of the cogito.

At the foundational point of the cogito, Descartes progressively deviates from reality. The fatal assumption Descartes made at this stage of developing his worldview was his interpretation that the cogito was the mind conjuring up ideas as the soul conductor of knowledge from intuition. His depiction of the experience is impractical; he isolated external sense experience and perpetuated redundant doubt in this critical tool for humankind's excavation of knowledge. If there can be no trust in the sense data of external stimuli, then anything outside the mind is meaningless. Descartes' categorization of sense data as uncredible isolates the mind from any valid reference point derived from the senses. Is the mind just inventing thirst? Do you not act in accordance with the senses? How is a skeptic like Descartes so certain that senses can only be confined to the external spatio-temporal context?

It is vital to doubt the senses with a healthy amount of skepticism because they can be wrong. But at some point, we must recognize that the highest form of knowledge rests in the sensuous detection of what is "most likely." The senses can be unreliable, but it is simply the unfortunate fact that sensuous detection is the absolute source of all things knowable. The mind and its senses are burdened, in the absolute sense, with the responsibility of formulating all meaning relative to all adversities brought about by its conditions. The composite of human senses and faculties of the mind stands as the initial cause; it is the utmost cause of badness, wrongness, ignorance, and evilness. It is much easier to see that philosophy's discrepant and muddied focus stems from the need for more systematized guidance for the mind.

It is important to note that I am not claiming that sensing is knowledge. Sensing is not knowledge, imagination is not knowledge, and intrinsic ideation is not knowledge. Rather, these are all critical human faculties responsible for human sense-making. The totality of faculties all carries importance in the construction of knowledge. These faculties of the mind work dynamically in conjunction with one another; they cooperate in such a way that conjures the function of synthesis. There is no inherent necessity to determine which human faculties are most or least important for synthesizing sensing into knowledge. We must construct meaning with acute consideration of the adversity given through the senses' fallibility. However, we must not deviate from reality as Descartes did at this point of reasoning. Cartesian skepticism went as far as to claim everything sensed could be put forth by some clever demon. However, what functional purpose does this serve humankind in the progression of constructing sound bodies of meaning? Cartesian skepticism eventually became the act of wearing out the logical "tool" of doubting.

One must avoid falling into the ad hominem thought process as Descartes did. Namely, the methodological source of knowledge universally determines the extent to which this knowledge is valid. Validity is not to be measured by the tool(s) used to synthesize data. Validity is measured by a proposition's ability to remain consistent with the environmental landscape it is embedded in.

VALID QUALITATIVE METRICAL ORIENTATION

"To him who looks upon the world rationally, the world in its turn presents a rational aspect. The relation is mutual."
—George Wilhelm Friedrich Hegel

Nothing would be as it is today without intricate methodologies of expanding our knowledge through the sensory mechanisms that drive scientific discovery. In the cogito, Descartes was attempting to optimize the human capacity to think well, but in attempting to build from it, he destructively diminished the value of the cogito by demonizing the senses. Descartes' premise critically excludes practical methods of empirical measurement and reliable testing methods relating to human environmental reference points. The senses are the foundational source for the astounding observable increase in the quality of human life. If we apply a guideline for a qualitative metric to judge the "truth" of his version of the cogito, this system renders his application of the cogito useless. The proposition of his understanding effectively distances his conclusions with a philosophy that can provide a more holistic view of truth.

Amid the conglomerate of humanly created ideas, science is rendered as the most successful. We must understand the features responsible for its success and what we can infer from these features to better orient the conscious mind with the context it is embedded in. Science's success can be attributed to three benchmarks that functionally provide helpful guidance in orienting the mind with its context. Firstly, content that results from the scientific method always provides fundamentally useful knowledge. **Secondly, theories that arise out of science parallel the measurably variable context of the ever-changing world with consistency.** A scientific theory can only exist if it is measurable, so it is technically never final. Therefore, it is always subject to changing congruently with the ever-dissipating natural world. Thirdly, it can generate causal models that provide knowledge of matters that **transcend our proximal field of perception**.

1. The idea of "usefulness" is relatively self-explanatory. The core of this benchmark is the consequential utility that results from any scientific undertaking. Usefulness can be defined as a consequence that leads to an effect of expanding consciousness. It remains consistent that any subject of scientific inquiry provides a certain value of desired utility for a purposeful end in the expansion of conscious reach. Suppose we compare the technological capacities between now and 100 years ago, the capacity to access information, access to

medicine, and relocate geographically are drastically different. All these outcomes produce a perceived benefit of expanding consciousness.

Usefulness can also be expanded past material utility. The mind's existential dread can be curbed (at least temporarily) by the scientific method's provision of purpose. If I am to begin the scientific process, I begin with a reasonable question that stems from the sensuous detection of a problem. For example, if I notice many people becoming rapidly ill in a particular location, I will use the scientific method to determine the causal model of the undesired result and reduce the cause of illness. Determining the causal model for this illness provides utility for the population affected and functionally sets a definitive purpose for my actions.

2. The ability to mirror the ever-changing parameters of the conscious environment is vital in consistently producing the value of "true." It is measurably true that the universe remains in a state of flux; therefore, our conclusions of truth must also be malleable to parallel the state of flux to reach success. The feature of science that gives it these qualities is the fact that all scientific conclusions remain falsifiable. Each claim discovered under this method always possesses the capacity to undergo infinite variations of testing and questioning. The mark of being able to withstand all tests allows us to assign a level of validity to

be assigned to a given claim. After establishing a high degree of validity through testing and questioning, we can continue forming meaning consistent with these measurements. Above all, the scientific method not only finds consistency with the ever-changing nature of our holistic environment, but the method also provides a rigorous system that measures the validity of a claim. Fluidity in the constant measurement of validity in any given claim allows it to be consistent with the fluidity of its context. Science's built-in adaptability is the pathway to any amount of certainty on knowledge in question.

     3. As established, the human senses are the foundation of everything we can identify as knowable. The five distinctly identifiable senses (sight, hearing, smell, taste, and touch) provide a proximal field of detectable material. Typically, our empirical observations are restricted to this set of detectable materials. However, through the expansion of technology and knowledge resulting from science, we have expanded the empirically observable field of detectable substances. For example, through the advancement of environmental sciences, we can predict what the weather will be like in the future. Without science, I could have never been able to detect that in 5 days, there will be a hurricane using my five senses alone. Through the scientific study of meteorology, I can know if a disastrous storm will occur in advance; with a proper warning, I can know if I should evacuate my home. The point here is that

through the scientific understanding of climate, I can obtain knowledge that transcends what my senses would have been able to predict otherwise. So when the hurricane hits, I will have access to sufficient knowledge to survive.

Indeed, the utility of expanding empirical detection beyond the standard proximal field of the senses merits the label "success." Of the senses we have transcended, it is vital to recognize that the particular sense of time should be recognized explicitly here. Science produces a causal model that predicts an entire temporal landscape that is, at least perceptually, unrevealed to the mind entirely. Yet, we have accomplished a model that can successfully predict perceivable events while transcending temporal bounds and simultaneously maintaining the property of empirical. In what universe is this not considered a success?

By extracting the benchmarks that give science its success, we can synthesize qualities of importance that allow us to form a qualitative metric. The qualitative metric will provide a tool capable of measuring the level of validity within additional qualitative material such as ideas, theories, and doctrines. As expressed, functional utility, infinite measurability, and expansion of detection beyond the proximal senses are all qualities that provide metrical validation in the claimed success of science. Of these three qualities listed above, functional utility and infinite measurability can be extracted as the two most applicable qualities for any theoretical or philosophical doctrine originating from the mind. If we utilize these two qualities as a qualitative system, we move toward a functional metric to assess the validity of a given philosophical doctrine.

The nature of this qualitative metric is emblematic of a pragmatist's philosophical view. Pragmatism measures the 'trueness' of knowledge based on the degree to which it is successful in useful application. Pragmatism is: "An approach that <u>assesses</u> the truth of meaning of theories or beliefs in terms of the success of their practical application. (Oxford Languages). This metric proves valuable because it holds any given concept accountable for being useful amid the extensive body of human knowledge and the environment. What is meant by "useful" here is that a given conception is serviceable for the expansion of consciousness that provides the consequence of human flourishing.

To exemplify this metric's functionality, I will introduce the problematic nature of the focus of solipsism. Solipsism is a theory of the mind and is relevant here because it can originate from the mental territory provided by Descartes' cogito. Solipsism is "the view or theory that the self is all that can be known to exist" (Oxford Languages). This philosophical doctrine is problematic because it lazily asserts that the only thing we can be certain of is conscious experience; therefore, it is all that exists. Even though the idea can be conceptually throttled in a way, this kind of reasoning results in tremendous problems in forming any functional truth. The framework of solipsism shares similar components to the position I am building. Like my position, solipsism stakes its base in a strong foundation, which is formed in the same cerebral space as the cogito: It can be determined that the thing we can be most certain of is thinking and being (consciousness); therefore, we can **only** be certain of thinking and being. In solipsism, this is used to assert that all there *is*, is your own experience, and nothing else can be certain.

A foundation is only a beginning; what sense does it make to determine the beginning of certainty as the end of certainty? Indeed, we can be **most** certain of consciousness existing, but what a lazy, arrogant, and extensive jump it is to assert experience is the **only** *thing* that exists. The interest of the qualitative metric is to discover *knowledge* that is both measurable and has the functional capacity to provide usefulness. If the experience of the mind is the only thing that exists, then all orientational components outside the mind that could serve as metrical components of reality bear no value in validating existential reality and therefore validating truth. Solipsism offers no useful way to orient the mind with any other truth relative to reality. In accepting solipsism, truth meets a dead end, morality meets a dead end, and value meets a dead end. If I accept solipsism, I accept that another mind's experience is an illusion. This opens the door to a purely dissociative dismissal of any material effect of my causal actions upon others that could quickly materialize harmfully. This offers no functional guidance in the context of existence nor increases the mind's capacity to "see" reality with any bit of purpose. Solipsism's foreground fosters a context that is exceptionally likely to produce an abundance of conflicting applications of meaning that are explicitly and starkly relative when operationalized. In doctrines like solipsism, truth takes an exclusively relative form dependent on the individual experience.

Above all, we can evaluate the critical example of pragmatism's capacity as a metric to steer the individual well in the landscape of such fundamental thought. The existence of the mind is indeed certain; however, certainty in essential meaning does not end with the certainty of the mind's existence; rather, it is where essential meaning begins. A pragmatic and scientific approach to investigating truth allows us to discover clues toward a path to meaning, orientation, and purpose. It proves as a great tool in helping us understand the importance of measuring the utility of an ideology, question, doctrine, or idea, especially in postulating validity and truth with certainty. If we wish for a doctrine, method, or postulation to be successful, we measure its capacity to mirror the elements that make science successful. If we do not, the chaotic fallout in truth and purpose will continue its dissipative course.

REBUILDING FROM THE COGITO

"You are an explorer, and you represent our species, and the greatest good you can do is to bring back a new idea, because our world is endangered by the absence of good ideas. Our world is in crisis because of the absence of consciousness."
–Terence McKenna

The aim of this text is to understand existence in a way that is nearly impossible to disparage. It is not enough to build a foundation; proper understanding entails accurately depicting the context that beholds the conscious mind. We have established the metrical components of utility, measurability, and consideration beyond the proximal senses. The steps that follow the cogito must remain disciplined to these metrical components and be procedurally consistent with how it was discovered. If the mechanism utilized to determine the utmost truth of the human condition is empirical, why would we stray away from it as we conceptualize truth further? And as mentioned above, the function of ensuring that ideas, theories, and doctrines are congruent with pragmatic qualities will steer us toward being correct. If we can use a pragmatically disciplined empirical observation to determine the utmost truth of the human condition, it must be the primary method of truth acquisition. Systematic guidance that is valid and consistent with the foundation and the whole of the greater human context must be applied to our pursuits of discovering deeper meaning.

The conscious observer must understand what determines their consciousness. If we remain disciplined to systematic empirical observation and sound reason, we can further determine that: since we are certain thinking and being exist, **thinking and being must have a cause**. The faculties that make up the conscious mind are products of the external spatio-temporal world. If the observation of the conscious experience is absolutely certain, its observational cause must be equally certain. And in the case of the human mind, the cause is inextricably embedded in its context. This context is encapsulated within the English language with words/phrases like "the external," "spatio-temporal context," and "nature." Reason permits us to build upon the cogito: "There is thinking and being (consciousness); therefore, there must be something that causes this thinking and being."

The causes we are searching for lie in deconstructing the constitutive elements of the environment that produce the conscious experience. This entails a methodical investigation into the foundational aspects of human knowledge, like language, culture, and biology. To observe consciousness in a more nuanced form, imagine a single glass of water. Think about the last time you drank from a glass. You can imagine a firm open-ended cylindrical barrier sustaining the fluid substance that sustains you.

The simple concept of a glass of water takes on a particularly bounded form through our limited experience with this functional object. But this item of perception, the glass of water, has much more depth to understand. Water is a fluid substance, offering a seemingly infinite potential to take on various chemical and geometrical forms. Yet, in this imagined concept of a "glass of water," the water temporarily holds the sustained and distinct geometrical form facilitated by the boundaries of the glass. In the literal sense, the glass is chemically compromised of sand, limestone, and sodium ash. In a phenomenological sense, the glass was formed through the confluence of heat, gravity, air, and artisan, giving the glass its essence. The convergence of all the assembled components of the glass leads to the consequence of the water's form. Though temporary, it is bound by the consequential events that synthesized the glass's form.

By understanding the components of the barriers that form the water, our conception of this "glass of water" becomes more apparent. Importantly, as the water in the glass has definitive barriers, so does human consciousness. Just as the barriers of the glass have a **cause**, so too does the barriers of human consciousness. An enormous set of definitive barriers categorically binds human consciousness. Understanding human consciousness, similar to how we understand the glass of water, will help contextualize it with the formative barriers that limit our ability to conceive of correct determinations of God.

In the glass of water metaphor, water represents consciousness in a particular way, a fluid and malleable substance with abundant potential. Conceivably, language, biology, culture, and many more elements comprise the conglomerate of barriers responsible for orchestrating the temporary, perceived form of any synthesized object. However, it is critical to understand that consciousness is distinct from water in that it can emanate acute awareness of its barriers **and** manipulate them. The ability to emanate awareness of the elements that hold determinant power over us is a critical component that will help us redefine our conceptions of deeper meaning and purpose. This concept is of utmost importance in understanding the central points within this text.

Many different doctrines are formulated in the same perceptual space as the cogito. Many attempts to orient the mind with what is *real* have yet to considered what orientation is most useful or consistent with existential reality. The phenomenon that we can deduce elements in nature that measurably play a part in **causing** our being is what renders doctrines like nihilism, solipsism, or simulation theory obsolete.

These three existential theories of reality explicitly state there is no use in exploring these causes. But if I can deduce that my conscious experience has measurable causes, why would I ignore these critical clues into my essence? The explicit lack of utility indicates that we can earn purpose by deducing important causal elements of our nature to deduce a true God. To do this, we must begin by critically analyzing what determines the perceptive appearance of what we are deducing from. Achieving a disciplined form of monitoring perception is no easy task but of utmost importance. Perception is the window into the empirical landscape that stands largely undetected as the mediator of what the mind understands as real. In other words, perception represents the most fundamental component of the barrier-forming glass which determines the shape of our water (conscious mind). To rebuild from the cogito in a disciplined empirical way, we must impose a critical evaluation of our perception of objects being perceived.

    Bertrand Russell performed a similar experiment to Descartes in formulating an exercise that allows us to determine what is knowable with absolute certainty. In *The Problems of Philosophy,* Russell introduces the exercise of considering the perception of a red chair. The vantage point from which you stand in relation to light in a given room will ultimately affect the shade of redness being perceived. If I stand on one side, with the light shining in the same direction as my gaze, the appearance of this chair will show the full and uniform shade of redness possessed by this chair. If I stand on the other side, and my vision is facing against the light, the chair will present shadows that may dim the shade of red or even make the chair's color inconclusive. This exercise allows us to distinguish between appearance and reality. Determining the true color of the chair based on perception is the essential nature of human responsibility toward asserting truth. The more variables and points of observation added to an object of perception, the more forms an object may inhabit.

Russell provides a different perspective from the cognitive space we find ourselves when we exercise doubt in absolute form. The fluidity of variables, capacities, parameters, and vantage points render the multidimensional synthesis between appearance and reality to be daunting. This can be discouraging because the act of perceiving is seemingly the most basic and self-evident starting point into which we assess truth. The appearance of any object beyond the self-evident cogito will feasibly take on an incomprehensible magnitude of forms. It is inherently contradictory for one thing to be many things; therefore, appearance is not a universal representation of true reality. Reality is the result of synthesis after a sufficient degree of appearance has been considered. Perpetual synthesis of appearance will eventually begin to formulate the condition of realness; but the bicameral assertion of real or not real can sometimes be a false dichotomy. The process of elimination is responsible for formulating that which deduces what is MOST real amid the fluidity of forms an object can take. By understanding the barrier-forming parameters of reality, we can begin to consider our causal relations of ourselves to this world. This level of consideration of our mechanisms of perception formulates primary orientational reference points that validate our sensuous understanding.

Bertrand Russell's exercise is essential here because the work done in this book effectively constructed a reality that entails meaning is only achieved through deducing the environment. Therefore, it is important to employ exercises that force us to critically evaluate the soundness of our perceptual processes if we are to rebuild a uniform conception that aligns with "true." Because our perception is an omnipresent lens that determines human understanding, we must understand that everything through this lens is one layer removed from reality. Thusly, the closest access we can get to reality is dependent on the extent of *appearance*.

In the following chapters, I examine distinct foci within the barriers of consciousness available through human perception, logic, and knowledge. A prerequisite to understanding the components relative to human experience is understanding concepts such as language, science, and phenomenology. From here, we can rebuild understanding from the cogito to reveal the additional parameters between perception and truth. The relationship between these foci is necessary to understand the dynamic relationship between language as a determinant barrier and the conscious mind.

UNDERSTANDING LANGUAGE AS A BARRIER

> "The world does not speak. Only we do. The world can, once we have programmed ourselves with a language, cause us to hold beliefs. But it cannot propose a language for us to speak. Only other human beings can do that."
> –Richard Rorty

Each component of this book is intended to construct a complete contextual awareness of conscious experience. At this stage in the book, we must progress our focus from understanding that reality is shielded by appearance and take on the responsibility of understanding what influences, determines, and manipulates the appearance of reality. This responsibility turns our attention toward the constitutive parts of "the environment." Language is wedged between appearance and reality, making it an omnipresent consequential component of the environment. Like perception, language is imperfect. As knowledge and meaning evolve, so must language. The proper application of Cartesian doubt brings the mind to a space where the very concept of language is deconstructed. At this point of the investigation, we must find a reason to believe that the foundational architecture of language provides measurable validity in its practice. In essence, absolute doubt entails the question: What renders the conceptual guidance of language certain?

There exists a multitude of functional theories of language and linguistics, all of which validate its substantive impact on collective human functioning. Universally, all thought must be assembled and communicated through the representational medium of language. Within this space of absolute doubt, there must be a deeper consideration of language's role in the formation of human meaning. The particular concern that arises from the space of absolute doubt is how language functionally impacts the manifestation of meaning at a foundational level.

The theory of linguistic determinism provides a useful lens for understanding the foundational impact language has on meaning. Linguistic determinism states: language directly determines how the mind conceptualizes and categorizes matter into meaning. This assertion suggests that, in practice, language is a determinant causal barrier forming categorical meaning. Language intimately impacts the foundation of meaning systems from their inception point. No matter what is being minded, language is the layered medium regulating one's capacity to comprehend any perceived item. Any given notion, concept, or idea entails representation by some form of linguistic symbolism. Furthermore, once a word's essence has cemented itself into being standardized, it profoundly impacts how we collectively internalize the subject matter a word represents. This assertion by linguistic determinism seems to be exceptionally apparent in the example of the word "nature."

Analysis of the word "nature" is useful in enhancing our understanding of the mind's relationship with language. "Nature" is held responsible for an extensive amount of explanatory responsibility. "Nature" is the linguistic representation of the holistic context encompassing consciousness. By understanding the content this word is designed to signify, we can unveil a greater degree of orientational understanding to be used in the formation of deeper meaning. **By understanding our mistakes in using the word "nature," a better understanding of the conscious mind's relationship with language becomes more accessible.** Western thought has not sufficiently grappled with the multitude of distinctive forms the word "nature" embodies. The inefficiency of Western thought and the English language regarding "nature" has led it to minding moral questions like: **If feasible, should humankind's moral code transcend what we understand to be the will of nature?**

Taking the above question to be of serious concern is symptomatic of flawed reasoning at the foundational point of human thought and wisdom. This relationship between humankind and nature is a fundamental building block of meaning we must get correct to provide accuracy at the inception point of any substantive view in the external and internal environment of the mind. Culturally, the word "nature" is unfairly designated to represent all barriers to the conscious mind in the active formulation of human understanding. Both understandings of "nature" can determine our contextual orientation at every moment. Given the pertinence of this word, it is imperative to ensure it represents the entirety of the concept it necessarily conveys in a definitive manner.

Is it valid to question whether our actions are determined by nature? Are we the hand of nature? Or are these arbitrary questions? Our inextricable connection to our environment entails that understanding our environment more accurately will allow us to understand ourselves. As deeper investigation uncovers, the standardized understanding of "nature" and "the natural" in the English language is comprehensively and problematically vague. The current definition of the word "nature" macroscopically misrepresents the essence of the material it is intended to communicate. Comprehension of this form of the word "nature" has led to the unfortunate effect of playing an integral role in perpetuating the result of dead ends in fundamental unsolved intellectual debates, as I will explore through examples below. Fundamentally insufficient definitions in any language can be critical bugs that lead to blind spots in understanding. The mind is responsible for creating clear awareness of this and modifying such tools that form its perceptual barriers.

It is critical to understand what we mean when we refer to the current standardized understanding of the word "nature." The word is defined in Oxford Languages as:

1. "The phenomena of the physical world collectively, including plants, animals, the landscape, and other features and products of the earth, as **opposed to humans or human creations**."
2. "The basic or inherent features of something, especially when seen as characteristic of it."
    a. "The innate or essential qualities or character of a person or animal."
        i. "Inborn or hereditary characteristics as an influence on or determinant of personality."

Each current use of the word presents its own set of issues. All the definitions establish a problematic, egocentric barrier between the human and everything else devoid of humanity. In fact, this mutual exclusivity is the primary misleading element within the definition of the word "nature." Consequently, this mutual exclusivity has been cemented through the perpetual application of its insufficient definitional construction. For instance, when I say, "I need to spend more time in nature," you may assume I mean to suggest I need to spend time in environments devoid of human intervention. Someone would likely assume I mean that I need to drive up a mountain and find some secluded outlook away from humanity. In formulating deeper meaning, like the idea of God, this orientational delineation between humanity and nature is mostly arbitrary.

    Critiquing the standardized understanding that "nature" entails exclusion from humanity is not meant to say there is no value in the conventional Western understanding. It is important to note that this understanding of mutual exclusivity between humankind and nature has some functional purpose. The orientational delineation between humanity and nature is valuable in differentiating between humanly edited genes and organically formed genes, identifying human and non-human carbon emissions, or even discerning between organic and artificial intelligence systems. However, nature's current conventional usage is disorienting and misleading in the larger context of holistically valid meaning-making in the greater context surrounding our species.

The construction of underlying exclusivity of the individual human vantage point within reality presents a blind spot when applying the word "nature" to form deeper meaning. This distinction is incredibly consequential in conceiving of the word nature in the application of doctrines that guide our morals and values. Considering this, the most accurate understanding of "nature" would be best measured by the benchmark of its practical use toward fundamentally orienting us with the measurable spatio-temporal landscape.

To further exemplify this issue with the definition of "nature," take particular lessons from *The Phenomenology of Spirit* by Georg Wilhelm Friedrich Hegel. In this text, Hegel proposes a universal method of reasoning he coined the dialectic. The profoundly integrated subject matter of this text provides a preliminary universal understanding of the human context. *The Phenomenology of Spirit* is widely regarded as explicitly metaphysical, but I don't suggest this to be the purpose of the text. Rather, it appears that the text was designed to reveal that the hazy presumptive label of "metaphysical" is mistakenly applied to concrete elements of reality that we haven't yet sufficiently detailed. The legendary "Hegelian dialectic" brought about by Hegel in this text presents a significant lens through which to study the essence of "universal." The Hegelian dialectic is a process presented as: thesis, antithesis, and synthesis. The assertion of the dialectic method by Hegel exhibits much value in illuminating a form of universal connectivity through a system of universal reason.

> Hegel proposes a critical and telling analogy that exemplifies the proposed dialectic: "The bud disappears when the blossom breaks through, and we might say that the former is refuted by the latter; in the same way when the fruit comes, the blossom may be explained to be a false form of the plant's existence, for the fruit appears as its true nature in place of the blossom." (1807, pg. iv)

There is an unfound beauty in this example that exemplifies the exact fluidity of *nature* and mind that I am trying to express. This unity is concealed by the structure of the contemporary definitions of the word "Nature." The dialectic is elaborated by the example of the 'natural' process of the bud blossoming into a flower. Within this analogy, there is a critical underlying substance that uncovers more about why the current understanding of the word "nature" is detrimental to the formation of deeper meaning. If we are to understand nature as "the phenomena of the physical world collectively, including plants, animals, the landscape, and other features and products of the earth, **as opposed to humans or human creations."** then the **process** of the mind and the blossoming of a flower is narrowly concluded as two utterances of distinct processes that are categorically distinct in a universal sense. Thus, the view of humans, as opposed to nature, creates this blind spot in understanding measurable connectivity that offers a complete appearance of reality as seen through the dialectic. As the bud represents the thesis, it stands in contrast to what it has yet to become, namely, the blossom. These two contrasting forms are necessarily reconciled (synthesis) to become a blossom. The reconciliation of notions within the mind shares this blueprint with the becoming of a flower.

As the mind is to discover truth, it follows a logical form that embodies the dialectical process; this is seen in the very nucleus of this text. The dialectic process seen here takes the form of philosophy as the thesis, science as the antithesis, and truth as the synthesis. The properties of science bring us measurable distinctions of causes, which distinguish the thesis as *narrowing* your perceptual scope. As will make more sense in the following chapters, the antithesis, science, stands as the contradicting form in that it reveals a teleological end to sequences of existence, and thus, it *expands* the scope. The synthesis of the most accurate truth becomes understood through the reconciliation of these two contradicting forms of conception.

In the first definition that asserts nature exists "opposed to humans," it is asserted that plants are considered natural;

therefore, they exist categorically opposed to humans. Now, imagine you could take your lungs and hold them in your hand. Without your lungs, you would simply cease to exist. So it follows logically that if a biological component of existence sustains human life, it is not opposed to your human form. Yet, if you could hold your lungs in your hand, the language used to describe them suggests a subtle but effective categorical distancing that determines them more as a biological item than an intricate fundamental component of your existence. In referencing lungs, the language typically used would distinguish them as "my lungs." Since you cannot exist without these lungs, would it not be more accurate if you referred to your lungs as "me?"

When we look at plants, we linguistically distance ourselves even further from this as a distinct biological item. Firstly, the definition of nature explicitly asserts that this given plant is not appropriately referred to as "me," given it is categorized as "opposed to humans" by definition. However, if we were to remove all oxygen-producing plants from our environment, there would be no oxygen to breathe, and our lungs would lose all capacity to sustain human life. Similarly, removing this item from the environment would cause us to no longer exist. According to the logical premise above (that if a biological component of existence sustains human life, it is not opposed to humans), plants cannot be considered as opposed to humans because they sustain human life to the same degree that our lungs do. Whether without oxygen-producing plants or our lungs, the consequence remains the same, we die. Therefore, the universal linguistic assertion determining the categorical delineation of nature and humans as distinct is untrue. Therefore, the first step in understanding *nature* is understanding that we are not distinct from it.

In any given language, there will come a set of words that offer explanatory power to a particular worldly substance that will come to an eventual dead end and become logically circular. This circularity arises at the edges of which human meaning realizes the extent of its substantive explanatory capacity. Typically, new terminology is coined as a mark of new concepts and ideas arising from the expansion of human knowledge. With the expansion of human knowledge, standing vernacular can realize a congruent expansion. The listed definitions in the second usage of the word "nature" above read as: "the innate or essential qualities or character of a person or animal. Inborn or hereditary characteristics as an influence on or determinant of personality." The problem of circularity is evident in the explanatory words in the definition of nature, like "innate," "essential," and "inborn."

Further analyzing the etymology, the Latin root of "nature," the predecessor "natura," originally asserts the description of qualities like birth, origin, and natural quality. All these words listed above are linguistic reference points for any concept that resembles an original cause. The problem here is that "inborn," "nature," and "innate" all similarly describe each other. This exemplifies a redundant trap of circularity which perpetuates a limited explanatory scope. The limiting suggestive scope and unnecessary circularity provide categorically insufficient definitional material that can be used. Circularity ought to be fixed if we wish to synthesize forms of holistic meaning with accuracy, especially in application toward the ideation of God. To avoid the unnecessary circular dead end, I suggest that the definition of "inborn" shouldn't share the same referential scope as "nature" or "innate."

The word "inborn" is straightforward; it should reference the determinant material of a human (or other living beings) from the origin of its vantage point. This entails that "inborn" refers to a form of original cause from the point at which the individual life began and the point at which the life ended. But it is the case that the inborn characteristics of a human being are not the singular original cause of nature because what is "inborn" entails its own set of causes that determined a being before it was born. The subject matter that the word "nature" identifies must be a holistic interpretation of what *is*. As scientific knowledge would assert, what *is* did not always entail the existence of human interpretation. Our language must have words that communicate this.

It is demonstrated in this second usage that the content "nature" is ultimately describing entails some description of *something's* determinant internal qualities. The assertion of "inborn or hereditary characteristics as an influence on or determinant of personality" seems to imply the inborn or hereditary composition of a being stands as the only set of determinants causes a single living being's existence. From the narrowed experiential vantage point of your life, this may be an appropriate application within given contexts. But to understand reality in an actual and holistic sense, you must consider that the very anatomy (mentally and physically) that determines my essence has its own set of determining factors. These determining factors reveal that our substance goes deeper than simply understanding an element of myself from birth. It is our responsibility to understand that fundamental "inborn" determinant elements of myself had their own determinants.

As mentioned, categorizing "inborn" with "nature" and "innate" offers a subtle poisonous impact on the scale on which we think. There is both a suggestion of limited temporal scope and a suggestion of restricted ascendent capacity. This particular flaw may be a result of the human perceptive condition, namely that our experiential knowledge of the world begins at birth. We do not immediately perceive the life of the creatures who came before us and shaped the very DNA we now possess. We must begin to piece together an expanded scope of linguistic consideration congruent with the modern body of knowledge. Mapping out the barriers placed on human cognitive potential allows us to initiate a line of thinking that transcends the limitations of our proximal senses.

The primary consequence of understanding "nature" in the same vein as "inborn" perpetuates the criminal tendency of seeing the world in a contextually limited temporal scope. Doctrines, qualitative theories, and philosophical work have been plagued by the circular suggestibility surrounding this form of understanding the word "nature." To provide an example from philosophy, *The Genealogy of Morals* displays "nature" in this limiting way. *The genealogy of Morals* was written by one of the most important figures responsible for advancing Western thought, Friedrich Nietzsche. I am not a Nietzschean scholar, so I will not be providing the utmost detailed account of all the points in *The Genealogy of Morals*, but I will give a summary of the most relevant claims and ideas that exemplify its scope as it draws distinct parallels to the linguistic scope of the word "nature." *The genealogy of morals* is a terrific, thoughtful, methodical, and deep consideration of the *nature* of human morals. Through his methodical investigation of the genealogy of human morals, a particular scope becomes apparent. In this essay, Nietzsche appears to be formulating, in an absolute sense, the *nature* of how morality is operationalized collectively by observing the known human history of relevant psychological tendencies relating to morality as a concept.

In this text, Nietzsche investigates, in the deepest sense, the essence of goodness and badness, moral adoption related to human tendencies of caste systems, and the depth and multitude of dimensions in perceived moral forms. There are significant inconsistencies between the grandiose essence of explanatory magnitude this book sets out to accomplish and the measurable selection of data used to execute this goal. Nietzsche remarks: "We are unknown, we knowers, ourselves to ourselves." (Nietzsche, 1998, Pg. 1). This is a central notion being broken down in my writing, but in Nietzsche's *Genealogy of Morals*, it doesn't fully belong. If we are to know ourselves, that will entail utilizing a scope encompassing all causal components that determine the composition that forms our self. As for *The Genealogy of Morals*, we see its scope of consideration plagued by the conventional linguistic suggestion influenced by understanding the word "nature" in the same vein as "inborn."

Understanding "Nature" and "inborn" in this vein would imply a rather specific temporal scope. Namely, a scope that exclusively considers the extent of a human lifetime. This vein leaves the mark of a stain in the very title of this philosophical text. To believe that understanding human *nature,* as it relates to morality, can be understood within the scope of human 'genealogy' alone is the mark of this collective misconception brought about by the boundaries set by the misleading understanding of "nature" as synonymous with "inborn." "Nature" is used forty-one times throughout this text; surprisingly, not all of them adhere to the assigned parameters of the limiting definition of this word. Throughout reading this text, Nietzsche extends the reach of his considered intellectual scope; he even investigates the role language plays in influencing our conceptions of the "good" and the "bad." I could feel his words trying to break through the limits of his language and tear open a greater path toward a more accurate truth. Near the end of this text, Nietzsche begins to consider matters associated with the original cause in the form of the Christian God and refers to *nature* in a seemingly Godly way by asking, "Why does *nature* give me a foot?" This usage begins to challenge language, thus initiating a suggestive wake in the coming stages of this word's evolution.

LATIN	LATIN	LATIN	OLD FRENCH	
nasci	nat-	natura		nature
	born	birth, nature, quality		*Middle English*

The etymology of this word features a recognizable evolution. The progression of the content this word describes doesn't yet entail a sufficient 'genealogy' of causes. We must take the responsibility to further evolve the word with our expanding knowledge of our existential context. By asking, "Why has nature given me a foot?" We open a wider breadth of consideration into the autonomy of nature, into the long-standing power it has had for eons before humankind, and into the cause of all causes that have determined our current state of being. This question begins to break through the narrow scope of time from the narrow temporal suggestions of "inborn." We begin to escape the temporal scope from birth to death to the truth of *nature*, the alpha and the omega.

The mistake Nietzsche makes is within his implied declaration of humankind's relationship to nature. The temporal scope of this deduction, as it relates to understanding nature, is the cause of insufficient answers in this writing and many others. If we take note of the scope he utilizes, it only considers a minute selection of interactive human data. The utilization of Nietzsche's set of data reflects the utilization of empirical reasoning, but the scope of the data doesn't sufficiently reflect the necessary context to determine such a relationship. This minute selection only contains the data from a select group of human societies dating as far back as the ancient Greeks in 480 B.C. This seems like a significant amount of time from the vantage point of one human being in one given moment in time. This notion ultimately aims to explain the context of humankind's relationship to nature. But humankind's relationship with nature began as far back as the inception of the causal mechanisms that brought about this Earth and everything originating out of it. Though Nietzsche's observation of patterns was genuinely astute, the data he used does not permit a judgment upon humankind's ultimate relationship with *nature*.

The essential composite elements of this existence are ingrained within the essence of the substance the word "nature" is designated to represent; this would entail a greater consideration of *time* is necessary. Nature's definition includes incomplete statements like "Basic or inherent features," "innate or essential qualities," and "inborn or hereditary characteristics." Each phrase lacks adequacy in translating the actuality of this word's necessary scope in both implicit and explicit terms. Remember, the task at hand is to build a sufficient theory of what/who God is; to do this, the implicit temporal suggestions in our language need to be addressed. The considered temporal scope of each definition presents a usage that is definitively applicable to the relatively minute time frame of a single human life. Yet, the substance of what is to be represented by this word was formed over the scope of all known time.

Innate, inborn, and inherent are not explicative of the word "nature" or its actual substance. When you consider the known universe, our systems of meaning demonstrate this capacity to be hyper-focused on the context of the temporal scope as it appears from a single human vantage point. Yet, there were moments in this existence when the human vantage point did not exist, and all there ever was, was everything as opposed to humans or human creations.

Due to the definition of nature being devoid of important temporal implications, it has failed to adequately orient the minds of those responsible for constructing substantial doctrines of meaning. The limited capacity of our lexicon has overwhelmingly plagued doctrines that attempt to make assertions on morality. In the case of *Genealogy of Morals*, Nietzsche asks questions in an astute effort to break out of linguistic bounds. But the parameters in which he operated still maintained a minute selection of temporal data. In the beginning, Nietzsche indicated that he planned to make us "knowers," known to ourselves. But to do that, we must start by considering a temporal scope encompassing a greater magnitude of informative data pertaining to the actuality of *nature*. It is true that the extent of human existence in *nature* is a minute fraction of time relative to the known temporal existence of *nature*. But not every item in the lexicon of our languages should be confined to defining this relatively minute time period, especially if the words can provide value in leading us toward a definition of God. By analyzing human genealogy, we can find some insight into our nature, but we will fall short of the whole truth without fully considering *nature* in actuality.

The standardized form of "nature" lacks the sufficient context-orienting substance that ought to be included within its definition. This deficiency can be observed in the doctrines dealing with forms of authority and power, as well as notations on political philosophy founded on certain moral convictions. Ideas with such explanatory responsibility subsequently fail because of the inevitable dependence on the standardized conceptualization of the human relationship to particulars in language that are poised to insufficiently interpret an adequate temporal context relating to the material it is designed to construe. The source of inadequate language can also be traced back to the capacities of our temporal perception. From the vantage point of a single human, the sensuous existence ranges from 0 to about 130 years. Our temporal sense provides data that is nothing relative to the temporal stance of *nature*. This moves the focus toward the next substantial barrier of the conscious mind, and we must actively work to understand and manage perception. A sufficient temporal scope is essential to forming deeper meaning, especially in the formative understanding of a concept like God.

Regarding linguistic determinism, inaccuracy in linguistic definitions consequently plays the role of a clever demon deceiving the collective understanding of human meaning. We can see this in our investigation of our relationship with the word "nature." By not perfecting the word that represents our environment, we are oblivious to the true context we are embedded in. Because language is an omnipresent element that actively determines the appearance of reality, misunderstanding any word can be consequential toward misguiding how the world appears to us. By understanding we are truly indistinct from nature, we open our consciousness to a greater breadth of compassion, responsibility, and potential.

SENSING, KNOWING, AND TIME

"The past is never dead. It's not even past."
–William Faulkner

As we continue to methodically break down humanity's contextual barriers around conscious awareness, you will notice that scientific knowledge remains a primary resource in reasoning. By understanding the world through the lens of pragmatism, utilizing scientific knowledge becomes necessary in formulating a holistic conception of the world. The efficacy of pragmatism's guidance can be further illustrated by investigating yet another barrier of consciousness that connects humanity to the natural world, the senses. Though sensing and knowing may be categorically different, the general extent of human sense profoundly limits knowing.

From the human vantage point, we can observe other species demonstrate limitations in knowledge caused by the limitation of senses. Because of this fact, it is only reasonable to conclude that the human species must also possess a limitation of knowledge because of a limitation in sensing. When we observe the evident extent of the temporal scope taken into consideration within a given doctrine, we are condemned to abide by the extent of how far the parameters of a relative scope seem to be. The reality of any given matter contains a relevant temporal context to consider, yet the appearance of this temporal context will rarely be revealed to the proximal senses. When developing a body of meaning that provides holistic explanatory power and more purposeful meaning, like God, time is the essential authority in validating it. Cross-culturally, God symbolizes the first cause and the final end; the explanation of God must include the temporal scope of the alpha and the omega. Ironically, this all-encompassing temporal scope gives religious dogma a monopolistic authority over common conceptions of God; they proclaim stories encompassing all time without a shred of substantive evidence. A breadth of myths and biblical stories are insufficient justifications for claiming a position of an all-encompassing scope of time. In developing an explanation for an empirical God, we must be able to answer the question: How can we determine an appropriate and sufficient temporal scope while maintaining a body of reason that is both empirical and certain?

The human mind's limited vantage point is a built-in deficiency in our environment that we must adapt to in constructing holistic bodies of meaning with certainty. Unfortunately for us, the empirical landscape only happens in the 'now,' yet even more challenging is that the appearance of 'now' can be argued to be an illusion because of temporal relativity. The events that appear to be happening now are not reflective of the actual time of the given event but the time in which we perceive it.

What I mean is, if I am sitting on my porch at night on my cell phone under the moon, the blue light from my phone, along with the light from the moon, will enter my perceptual field 'now,' but the event of light reflecting off the moon and into my eyes happened 1.3 seconds before my phone light happened. So, I cannot say both happened at the same time even though they are both happening to me now, at least according to my proximal senses. Likewise, the pain I feel if I hit my head will be felt before the pain I feel when jamming my toe. The point here is that there exists an illusion of "now," yet our immediate proximal senses don't detect "now" with proper contextual temporal accuracy. Lack of accurate temporal awareness illuminates a challenge in which the intellect is responsible for navigating while configuring holistic structures of meaning regarding time.

The implicated challenge here is that eons of consequential dissipative events of *nature* have caused the aggregation of perceived data (appearance) we take in. Concerning certainty, our temporal condition requires us to begin by asking: How can I be certain the past is even real? How can I be certain the future is coming? If our perception is limited to the present moment, how can the past even be empirical beyond this moment? The necessity of these questions illustrates the damning adversity of the conscious mind in creating grand conceptual ideas. The senses are the only thing we possess to detect substance; the capacities of our intellect stand as the only mechanism which allows us to synthesize realness beyond the appearance of the moment. Luckily, the sensuous experience of the conscious mind is not completely isolated within the proximal temporal reach of the moment it perceives. Memories present a simulation of an internally sensed experience from a prior experience that allows you to peer beyond the given moment. A memory may seem like an obvious phenomenon that doesn't need explanation or validation, but it is critical to remain disciplined in a consistent analysis of the senses.

      The experience of memory bears many similarities to the projected experience of the cogito. Memories can be empirical if it is synchronously verified as consistent with external metric points. Being certain of the past requires various forms of experimentation; through measurability, we can achieve a high degree of certainty. For example, if I take a tennis ball, hold it out in front of me at shoulder height, then let it go, there is no suspense in wondering what will happen next. We understand from our working memory of observing the physical nature of our existence that gravity will cause the tennis ball to fall. In this regard, we can be certain that the appearance of the past is real because of its measurable success in predicting the results of events that have yet to happen.

Envisioning a tennis ball dropping is an elementary example, but it demonstrates our propensity to empirically assemble a multitude of clues that allows us to formulate causal models that successfully predict the existence of a past and a future on either side of the temporal frame we perceive. Systematically guided empirical mechanisms allow us to expand the scope beyond our proximal temporal perception. Most importantly, if we can validate that the past is certain, we unlock validity in utilizing vast theoretical knowledge. More specifically, we can use the theory of evolution as a valid contextual reference for understanding our existential vantage point.

Understanding how time affects the qualitative vantage point in the observation of meaning is critical in contextualizing what the conscious mind is from the limited perspective of the conscious mind. Reducing or expanding a given amount of time considered when focusing on a particular notion will ultimately change our conceptualization of the event. The point is that the human detection of time is a critical determining factor in the resulting notion from the synthesis of any conceptualization of reality. The temporal condition of the mind has often served as the ventriloquists of deception hindering the succession of Western ideas. I propose that this sensory disposition relating to time is the frequent cause of the "insolvability" in particular philosophical problems. For example, take the opposing epistemological theories of rationalism and empiricism. Rationalism and empiricism are philosophical theories positing mutually exclusive understandings of how we proceed to *know*. The perpetuated mutual exclusivity assigned to these doctrines illustrates the human mind's propensity to dualistically categorize abstractions excessively.

"Rationalism" is defined in Oxford Languages as "The theory that reason, rather than experience, is the foundation of certainty in knowledge." If we are to accept this doctrine exclusively, it will entail the acceptance that knowledge is a result of innateness, reason, and further induction. Rationalists deny that sense-experience is a reliable method to conclude any form of knowing. Descartes' reasoning after the cogito is widely regarded as responsible for the paradigmatic advancement of rationalist philosophy.

Conversely, empiricism is "the theory that all knowledge is derived from sense-experience. Stimulated by the rise of experimental science." (Oxford Languages). This doctrine asserts that sense experience is quite literally the only functional way to obtain any bit of certain knowledge. Empiricism denies that knowledge can result from innateness, reason, and further induction in the formation of knowledge. It is widely known that

John Locke, George Berkeley, and David Hume are responsible for perpetuating these notions that the senses are the best path to knowledge. Many argue that the pragmatist lens limits us to accepting empiricism exclusively, but I argue that the pragmatist lens, congruent with modern scientific knowledge, would rather expand the way we think about the nature of the argument between these two theories of knowledge. Thus, the rationalist and empiricist argument can be dismissed as a false dichotomy of the past.

These rules assigned to the universal epistemological assertions described by each doctrine are rigidly antithetical. Yet each of these doctrines can functionally explain observable instances of knowledge acquisition. For example, how could it be the case that a baby innately possesses instinctual knowledge of how to consume milk from its mother upon birth? How would that same baby know not to touch a burning hot stove without the sense-experience of touching it? If rationalism were true, what would be the point of academic institutions that formally transmit knowledge by reaching minds through the senses? In a classroom, the student must audibly and visually detect external stimuli, then further synthesize them to acquire the knowledge being transmitted. Certainly, neither of these doctrines is universally true, and neither remains universally consistent with all known examples of knowledge acquisition. By referencing each doctrine's application in the context of the mind and its relevant spatio-temporal context, we can understand that each bears some reflection of truth, at least in anecdotes.

With each doctrine's proposal, philosophy asks: "Which is true?" The primary issue in answering this question has been that all argumentation from each side has utilized explanations and examples that don't entail a sufficient temporal scope. Expanding the temporal context taken into consideration will allow you to appropriately analyze this profound disagreement, thus justifying the notion that the two form a false dichotomy. I'm aware of the sophistication of the back-and-forth logical argumentation between rationalism and empiricism over the last few centuries. But the circularity of this philosophical debate suggests that philosophy must utilize additional tools to dilate its focus and realize the value of applied evolutionary biology.

It is important to note that my utilization of the theory of evolution doesn't exterminate the mystical, esoteric, and wondrous nature of knowledge, consciousness, or purpose. The theory of evolution offers the best possible guidance in clarifying a proper theory for the phenomenon of knowledge acquisition. To better understand why the theory of evolution helps clear the muddy waters of the disagreement between empiricism and rationalism, it is helpful to consider the concept of instincts.

An instinct is a non-learned behavior that a living being will enact to perpetuate its survival. Because it is an applied behavior that serves an important function, an instinct must be considered an object of knowledge. The tricky thing for philosophy is that it doesn't perfectly fit into rationalism or empiricism. An instinct is an applied behavior that any given species on this earth will demonstrate at some point. An excellent example for understanding the concept of instincts as a form of applied knowledge would be the story of the beaver.

The beaver comes from a species genus known as the Castoridae, the precursor to the modern-day beaver known as the castor Canadensis. This interspecies development is estimated to have taken course over 15 million years, approximately 33 million years ago. The Castoridae, now extinct, was a large rodent best defined as a cross between a gopher and a bear. During the long process of its extinction, its environment experienced harsh and abrupt changes. Extreme climate change fostered a new harsh, cold, and dry environment, leading to a perfect storm of complex challenges that required extreme adaptation. Because of this climate change, their environment contained less hiding space, less prey, and new predators. During this phase, the only known adaptive method they had developed for survival was to eat wood-based plants and hide underground. But this strategy proved ineffective, and they slowly died out over time.

At one point, a clever Castoridae rodent with an exceptionally high level of neuroplasticity and intelligence figured out how to combine wood and water to optimize its chances of survival. Using tree branches to build secure and safe structures in deep pools of water, it acquired the ability to build dams. Building these structures resulted in a new chance of survival; the rewarding nature of this behavior caused it to propagate and be transmitted to further generations. After many generations of these rodents practicing dam building and living semi-aquatic lifestyles, the castoridae's genes began to mutate, which resulted in the formation of a flat tail, webbed feet, and an unprecedented ability to hold its breath underwater. The decisive and clever problem-solving of one Castoridae functionally changed the environment through adaptive behavior. Incredibly, this led to the genetic development of an entirely new species.

The beaver example sheds light on the temporal consideration left out of the disagreement between empiricism and rationalism. The synthesis and subsequent action consequently caused the Castoridae's genetic sequence and transmitted knowledge to form. It has been observed that dam building has become so ingrained in a beaver's behavior that all they need is to hear running water, and they automatically start building a dam. Here we see a remarkable example of a given creature's learned, adapted behavior being the genesis of new genetic inheritance and newly engrained knowledge. The important implication here is that epistemological inheritance had a point of inception. As we expand the temporal scope of information from our vantage point, we see in the instance of the beaver that the innate knowledge of an instinct was formed out of synthesized sense-experience 33 million years ago.

It is important not to perceive this example as a complete dismissal of rationalism. It appears that the genesis of this knowledge resulted from sense-experience of the beaver's environment. However, at the inception point of newfound bits of applied knowledge, like a newly developed instinct, a synthesis of mind generates novel material for the world. In neuroscience, this occurs when a measurably large amount of neuroplasticity in the mind fosters new creative solutions to novel environmental changes. The important note from this example remains that both epistemological theories are incomplete by themselves, and the delineation between them is arbitrary. This is not to prolong the argument between the two theories but to say that this argument is a waste of time. I argue that neither empiricism nor rationalism comprehensively explains knowledge acquisition, but they are two essential components to the mechanism from which we come to obtain and create knowledge. It will always be true that the inception point of every concept will have been derived from some way of sensing. However, sensing is not knowledge, and there is always a synthesis involved, especially in creating things like language, math, and theories of reality. To remain disciplined to the principles established in this book, we must accumulate more examples of scientific knowledge we can further synthesize to form a better-contextualized understanding of a complete epistemological theory that remains congruent with reality.

Nonetheless, it is time to do away with the rigid assertions of doctrines that contradict one another and bear inconsistent metrical contingency with the environment. Debating the origin of knowledge utilizing data within the proximally detectible field has produced minimal utility in orienting the mind with its experiential context. We should no longer seek knowledge of original forms through direct analysis of data limited to the proximal senses. The advancement of the theory of evolution introduced human perception to a monumental expansion of an observable causal model of human genesis that dwarfs any genealogical record. Evaluating the world through the lens of evolution provides a tool that allows us to adapt to our limited sense of time to formulate a contextually sound conception of more profound meaning. The theory of evolution is the best resource for understanding causal models that predict and explain objects of reality beyond the perceptual temporal reach of the conscious human being.

EVOLUTION AND CONTEXT

"Nothing in biology makes sense except in light of evolution."
-Theodosius Dobzhansky

By creating causal models through the scientific method, we have expanded the capacity of knowledge well beyond our proximal senses. Thus, we have developed a method to sense what was formerly unsensible. In essence, science is our 6th sense, and it is the method by which we expand our environment and, consequently, necessary behavioral adaptations. Science is the tool that will allow us to excavate the elements of knowable reality for the necessary purpose of constructing a temporal scope that may include the alpha and the omega. By denying this fact, you commit yourself to a lifetime of dogmatic drowse and false enlightenment.

The logical progression of the book since we began *rebuilding from the cogito* follows: to orient ourselves with reality, we need to understand ourselves, to understand ourselves, we need to understand our environment, to understand our environment, we must understand nature, to understand nature, we must understand evolution in a more nuanced form. Our cultural understanding of biology has been particularly reductive in isolating mechanistic procedures rather than creating critical awareness of biology's implications toward our essence. Very seldom is it apparent that empirical observation provides us with the critical answers of "why?" However, any given data set contains an insurmountable collection of additional relevant material embedded within it. No matter what substance is being perceived, there is always more to it than meets the eye. It's important not to be too imaginative when deriving knowledge supplementary to a given observation. However, a disciplined imagination gives birth to a significant breadth of insight. Never fall for the notion that just because truth is dependent on the circumstantial determination of what is observably true, it can be mundane. Nothing is mundane or banal; the projection of values can only be cast out from the eye of the beholder who doesn't understand that substantial knowledge can be derived from what can be sensed. The world is a mirror in this way: the projected mundane only returns from where it was projected.

    A deeper understanding of the scientific theory of evolution brings us closer to understanding the context of our conscious being. Remember, nature must not be confined to the definitional understanding that humans are distinct from it. Though our understanding of evolution is incomplete, the theory of evolution provides a causal explanatory model that entails an extensive temporal scope while remaining empirical. As most know, the theory of evolution was proposed by Charles Darwin; it generally postulates that as species interact with the environment, they acquire mutative traits that help them better survive within the environment. It has been observed that as a mutation finds success in the perpetual application, it is then passed down to the following generations. The traits that evolve from these environmentally reactive mutations take the form of both behaviorally and physically inherited traits.

A famous example of the evolutionary process is Darwin's observation of different types of finches on the Galapagos Islands. In his observations, he encountered a multitude of finch species variations that differed from island to island within the Archipelago. Darwin postulated the biological variations between the different species of finches varied with the food source on each island. If a food source on an island happened to be large nuts contained in shells, the finch's beak was observably more robust and pointier, so it could break the nuts open more effectively. If the food source on the island were fruit, the beaks would look more rounded, like a parrot's beak. If an island contained more insects to eat, the beaks were narrower and pointier so they could pry insects out of small hiding spaces. These observations revealed that an ancestral finch whose genetics were exposed to new requirements for survival mutated new traits based on their behavioral efforts to combat these challenges. The process of evolution is universally observable in any creature that demonstrates some form of interactive agency with the environment. The empirical revelation of evolution opened a new window into the process of life. We have come to find it is the mechanism that brought us to the human experience as we know it.

The finch example is a common introductory lesson to understanding the theory of evolution. However, this example is not optimal for truly understanding evolution. The finches on the Galapagos Islands don't seem to reflect a high degree of agency, as seen in the example brought forward by the beaver story. The finch example doesn't explicitly demonstrate a creative synthesis from an evolving being. In the finch example, it is implicit that the bird species was seemingly blown around by the wind and thrown into an environment where there is only one food option that its genetic makeup autonomously adapted to. The point here is that evolution is much deeper than a simple mechanistic procedure of genetic adaptation. There is a much deeper interactive system at play, one where consciousness plays a much more significant role than the standardized biological understanding would suggest.

Since the inception of the theory of evolution brought forward by Charles Darwin, we have seen an advancement of definitions of this theory and the development of many sub-theories that affirm Darwin's initial findings. More astutely distilled definitions of evolution include: "change in the properties of groups of organisms over the course of generations…it embraces everything from slight changes in the proportions of different forms of a gene within a population to the alterations that led from the earliest organism to dinosaurs, bees, oaks, and humans." (Futuyma, 2005: 2). In Futuyma's definition, notice the language alluding to omnipresence in the phrase "it embraces everything." Another helpful definition is: "Evolution may be defined as any net directional change or any cumulative change in the characteristics of organisms or populations over many generations—in other words, descent with modification... It explicitly includes the origin as well as the spread of alleles, variants, trait values, or character states." (Endler, 1986: 5). Regardless of which definition you consider, the fact remains that this is the process that demonstrates the inception of being, which takes place out of our proximal sense range, yet is omnipresent.

    The entire timeline of the 2.23 million known species is still a puzzle yet to be solved with acute certainty. However, by analyzing what we know at the base of all living animal organisms, we can broadly construct basic universal mechanisms of evolution related to the environment. Of all known species, a particular organism is regarded as the first animal in the ancestral family tree of animal life. This primordial ancestor also marks the genesis of the nervous system, which has inaugurated the reveal of a timeless dialectic in nature.

This particular species is formally known as the sponge. This organism has a coral-like appearance that could cause someone to confuse it with a plant, but several features qualify it as an animal. It has a series of dermic layers comprising a multicellular system that helps it consume and digest organic oceanic substances like detritus, plankton, viruses, and bacteria. They interact with the environment through consumption and produce oxygen and collagen back into the ocean, providing rich elements for other marine life. The sponge is also important because it is believed to be the first identifiable species with a rudimentary nervous system. A team of researchers recently discovered sponge cells that mimic the composition and function of the nerve cell synapse in humans (Musser, et al.). This provides strong evidence that these sponge cells provide the original framework of the nervous system from which human cells evolved. Evidence suggests that sponge cells are the bridge leading to sensory motor navigation of the natural environment.

The nervous system is a complex network of cells and tissue that coordinates physical function and detection within an organism's environment. It is the central operating component responsible for sensing and responding to environmental changes, controlling movements, regulating organs, and processing information. Establishing a general timeline of the nervous system is critical because the nervous system is the primary evolutionary component interacting with the environment that humans can sensuously detect.

Evaluating ancestral species from the sponge can provide profound insight into fundamental universalities in the evolutionary process and materialize Hegel's famous dialectic. Following the sponge, two notable evolutionary successors are the hydra, then the flatworm. The hydra is a transparent-bodied animal with a stem for a body and appendages with a plant-like assembly. It dwells on the ocean floor, appearing like a micro-plant but isn't rooted into the ground, its motion is limited, but it can summersault across the ocean floor. The hydra is known to have the simplest form of a nervous system comprising hundreds of nerve cells. The nervous system can detect the mechanical movement of prey and light as it has reactionary appendage movement to each environmental stimulus. You could imagine this as an unrooted sponge with measurable sense detection allowing it to hunt tiny prey.

Further down the evolutionary chain, we can observe the emergence of flatworms. Flatworms undoubtedly have a more developed nervous system when compared to the hydra, but the extent of what each can do is comparably similar relative to the human vantage point. Flatworms do observably have an extra sense of vision, as they possess small eyeballs on either side of their rectangular bodies. However, the flatworm's eyes primarily detect light as the hydra's appendages already do. The real discrepancy between the hydra and the flatworm is in the flatworms' vastly greater capacity for mobility. The flatworm is a simple creature, but its bodily structure allows it to be one of the first ocean animals with the ability to swim. This organism's physical assembly allows it to move beyond what it can even sense. This illuminates the critical takeaway in the evolutionary relationship between a given being's environmental mobility and sensuous capacity. It could be that the hydra's overshadowing capacity to sense its environment with little to no mobility prompted the nervous system's expansion in capacity to mobility within the spatiotemporal environment, thus stimulating the evolution of a new species with more mobile capacity. This relationship provides explanatory power in expressing the determinant benchmarks of evolutionary indicators within a living being that signal the need to expand biological capacity. This capacity to sense the spatiotemporal environment and move in the spatiotemporal environment are such indicative benchmarks. When one capacity outmatches the other, the other is forced to make the next adaptation to "catch up." This formula presents an essential dialectic within living organisms relating to the environment. As we observe the function of this process, we see a materialization of the Hegelian dialectic at the basis of our most fundamentally innate nature.

The sensuous capacity of all mobile living creatures adheres to the functional extent of their given nervous system, which is the mediator between sense and mobility. Zooming out, the implication of a given living creature's nervous system and consequent existential understanding is condemned to be acutely specific to its environment. All animals (including humans) have an evolved sensorimotor nervous system that directly reflects the environmental demands in a spatiotemporal fashion. The history of a creature's engagement with the spatiotemporal environment will indicate the extent of its capacities. The dialectic within a living organism would be an independent actor that makes up the broader evolutionary dialectic between an organism and the environment, which is the primary essential dialectic of nature.

This brings the focus back to the current state of evolution in humanity. Humanity hasn't evolved the utmost keen proximal sense capacity when comparing human vision to an eagle or sense of smell to a dog. However, creativity has set our species apart because it allows us to manipulate our environment better than any other creature. Humanity is the most complex manifestation of a being who creates within observable existence. This observable element of creation exemplifies an advanced ability to exercise awareness and synthesize sensed information in its environment with an observably high level of neuroplasticity. This adds a particular degree of nuance when trying to understand the manifestation of evolution in humans. Throughout human history, we have exponentially increased our ability to determine our environment, thus placing the process of evolution in unprecedented territory. With this nuance in mind, dual inheritance theory offers the best guidance in understanding the human environmental context in the pursuit of deeper meaning.

Dual inheritance theory posits that humanity's biological inheritance is no longer limited to genetic inheritance. The theory distinguishes between a genetic line of inheritance from your biological parents and a cultural line of inheritance from other members of a given cultural society. Dual inheritance theory explains why we are seeing exponential increases in technology, deviations of social norms, advancement of communicative systems, and grander universally explanatory doctrines. But more importantly, the effect of this humanly created line of inheritance is that it increases our ascendency over our environment and, thus, genetic inheritance. Our increased ascendency becomes apparent when we analyze the history of traceable cultural advances that reflect changes in our genetic inheritance.

As we study human history, we can observe that specific cultural habits influence specific phenotypic and genotypic advancements in human genetics. For example, the creation of tools, fire, and efficient hunting methods provided early humans with more time and energy to develop organized cognitive skills and methods to re-transmit these learned environmental solutions. The time to develop symbolic communication is the likely cause for the development of cortical regions of the brain devoted to language and symbols. Furthermore, the invention of portable water containers caused particular human groups to be more mobile, thus causing the development of higher arches in the foot. On the other hand, the generational transmission of specific tools and methods for shelter caused our bones, muscles, and tolerance for harsh climates to weaken. From this, we see that evolution doesn't universally entail improvement. Evolution entails survival-based adaptations in accordance with the environment. So, when the environment doesn't provide harsh challenges, our genetics will adjust accordingly. This places a particular responsibility on all humanity to understand the evolutionary consequences of our behaviors critically.

Dual inheritance theory doesn't demonstrate a deviation from universal historical patterns of evolution but affirms it. This theory affirms the primary dialectic between a living creature's conscious management of genetic inheritance and the existential environment. Behavior has always been a major player in manipulating a given species' environmental engagement and, thus, inherited essence. The exception is that humanity has created a social network of intergenerationally transmitted behavior that consequently determines long-term genetic inheritance. Through symbolism, language, and technology, humanity has manufactured an additional line of encoded information that determines the future of the human genome.

By manufacturing the sixth sense, an additional line of inheritance, and the ability to apply and synthesize information to expand our technological capacities, we have exceedingly gained immense control over our environment. The historically autonomous omnipresent process of creation is now becoming disproportionately controlled by the subject we consider to be conscious. We are not the only creature to manipulate our environment, but evolution has not yet seen this degree of consequential environmental manipulation humanity displays. We have manufactured greater mobility in expanding the sensuous intake of information far beyond the proximal sensuous reach. Thus, we exercise "senses" not produced solely by the physical senses but by our ability to imagine temporal and spatial projections. The environment created by our actions has removed much of the harsh physical elements of our environment; however, the environment we are actively creating fosters complex challenges for the intellect. If the dialectic of the nervous system is true, our extraordinary ability of "mental sensing" would suggest that the next stage is the evolution of greater "mental mobility" instead of physical.

Dual inheritance theory is incredibly helpful in contextualizing us with the causal mechanisms of our more recent past. In the next chapter, I will elaborate further on why it is such an essential theoretical framework for understanding deeper meaning concerning existential purpose. Understanding the nature of our contextual being through dual inheritance theory is like utilizing a telescope to explore outer space. It is an incredible tool for understanding truth in our origins and fate, an essential quality the idea of God represents.

EVOLUTION OF CONSCIOUSNESS

"Ignorance more frequently begets confidence than does knowledge: it is those who know little, not those who know much, who so positively assert that this or that problem will never be solved by science."
–Charles Darwin

Consciousness and the display of conscious behavior play a vital role in the formation of life. Understanding the reality of consciousness this way is fundamentally important in proper contextual orientation with human reality. The environment that a living creature engages with is the determinant factor of its essence. However, the conscious mind has an omnipresent variable level of ascendency in determining what element of the environment it engages with. Interestingly, this entails that a creature's conscious awareness operates like the primary component of its environment. Meaning if I synthesize information from my environment to conjure up a notion or idea, then that notion or idea is now considered a working element of my environment.

This entails that human consciousness is the primary determinant of the environment, which determines its inherited genetic composition. Now, when we observe the "consciousness" of a hydra, its sensuous capacity only allows it to detect motion, light, and food. Therefore, its extent of possibilities doesn't give its consciousness many variables to sift through. But as we move down the evolutionary chain, from the flatworm and beyond, a pattern of increased sensuous and mobile capacity becomes abundantly clear. The primary driving force of this increased capacity is consciousness' inauguration of synthesized elements, which emanate the awareness that a greater capacity must be obtained. Thus, the lineage of species acutely displays growth in consciousness, conscious capacity, and conscious awareness as the evolutionary timeline unfolds. Through a given being's emanation of a more holistic environmental awareness, consciousness begins to realize itself, to itself, and of itself.

Theodosius Dobzhansky's words: "Nothing in biology makes sense except in the light of evolution," reign true not only when applied to biology but to philosophy. As distinguished earlier, philosophy is hard to define, but its purpose must be to offer existential guidance in light of the overwhelming exponential increase in technological power. The guidance offered by Western philosophy is often quite ambiguous or entirely unhelpful in this regard. However, a consistent theme within many of the major questions asked by Western philosophy offers a rather unintentional line of guidance. Philosophy asks questions until they become unanswerable.

The benchmark of insolvability can serve to indicate that we have come to the end of our sensuous reach. As established, leveraging what we understand from the theory of evolution unveils knowledge that was formerly unsensible. As the environment of the human being consequently expands, our intellects bear more responsibility to organize how the world is understood. This immense responsibility is where disciplined methods and rules of thinking keep the mind in realms of rationality and away from far-fetched imaginative and dissociative ideas. However, with an expanding body of knowledge, no rule should ever stand unquestioned or untested.

In humanity, the evolutionary process has been hijacked by the conscious mind, whether the mind is conscious of it or not. Our conscious intervention may cause yet another turn in the evolutionary process into the advancement of the metaphysical or, more specifically, transcendent consciousness. When we consider the notion of "consciousness," we see that philosophy hasn't reached a consensus on what the word definitively means. Yet, consciousness is an omnipresent element within the sensuous gaze. From the human vantage point, philosophical consideration is missing eons of experiential sense data relating to consciousness. You will find that it is indeed advantageous to understand consciousness analogous to a glass of water. In this analogy, the water represents consciousness as a shapeless substance, whereas the glass represents the holistic environmental parameters that define it (language, knowledge, sense capacity, habits, epigenetics, etc.). With this imagined concept in mind, understanding a more contextually detailed reality is more feasible.

From human to human, each environment will offer slight variations. Suppose we are to picture each individual's environment portrayed as each individual glass holding its respective volume of water. In this imagery, we can feasibly picture slight variations of glass sizes and shapes actively determining the condition of the water it holds. The variability of each environment represented by each glass would be greater with differentiation in age, time period, culture, climate, socioeconomic status, and other variables. You could also imagine that the spectrum of different holistic environments would extend to species beyond humanity. Imagining variations of consciousness in different species would be analogous to envisioning glasses of water next to plastic petri dishes or bowls made of clay. Thus, each environmental barrier fluidly expands or contracts the condition of the water.

The only thing the "glass of water" metaphor doesn't convey is that consciousness has always been a relevant player in the history of evolution. Understanding consciousness better would require a deeper investigation of how consciousness can be measured through the history of species leading up to the human. I brought about examples of consciousness in the cogito and mere statements alluding to this idea, but here, it must be more definitively defined with the proper scope established. Current understandings and definitions of consciousness are emblematic of the human mind's propensity to be far too reductive. It is often either reduced to a mechanistic process of components within the brain or some mystical, magical force.

Consciousness entails an esoteric sentiment from the human observer, but there is much value in being able to discern what measurable elements indicate and contextualize consciousness in a way that diminishes any element of mystery. In philosophy, consciousness has been referred to with terms like "phenomenological," "experience," "awareness," and "subjective." But perhaps the most useful reference to consciousness comes in the form of a question proposed by Thomas Nagel, which goes as follows: "What is it like to be a bat?" Nagel proceeds to define consciousness by stating: "An organism has conscious mental states if, and only if, there is something that it is like to be that organism." And through this understanding, we have guidance in understanding subjective conscious experience that remains congruent with the proper understanding of "nature" and temporal scope.

Many philosophers have made the parochial claim that humans are distinct from animals and nature in that we possess consciousness. Through the lens of evolution, this distinction cannot be possible. Curiously, the primary example used to distinguish animal behavior from human behavior is describing animals as reflexive and instinctive, thus not containing mental sequences humans perform. However, the example of the beaver's instinctual ability to build a dam displays how this instinctual knowledge was conceived in the original synthesis of the mind. The last Castoridae manufactured a new environment by demonstrating awareness and ascendency within their environment. This awareness and synthesis formulated a new semi-aquatic life which was the direct cause of mutation in their genes over the long duration of time spent in the new semi-aquatic environment.

The beaver example suggests an implication pertaining to the role of consciousness in life through evolutionary understanding. Might contextual awareness and subsequent creation be measures of consciousness? Consciousness is defined in the dictionary as "the state of being awake and aware of one's surroundings. The fact of awareness by the mind of itself and the world" (Oxford Languages). This awareness of surroundings was demonstrated in the early Castoridae rodents 33 million years ago. We also see this capacity to emanate awareness of its surroundings and create its self-conscious environment within its given environment. This ascendency is something that they show to have in common with humans. In this example, the beaver's behavioral adaptations materialize as evidence of consciousness in an earlier form.

We can define consciousness as "awareness of your surroundings, the ability to create, awareness by the mind of itself, and imagined existential experience." When we define consciousness with these terms, it may seem like we will have trouble using the Castoridae/beaver example to justify "awareness by the mind of itself." In fact, that has always been a critical issue in determining animal consciousness. We can justify that it was aware of its surroundings, we can definitively say it can create, and we can imagine what it is to be a Castoridae or beaver. From our own vantage point, we can definitively say that self-realization is an undeniable part of being a human, but we cannot speak on behalf of other creatures in this regard. We cannot become a beaver or recreate its experience with certain accuracy. However, a resource that provides insight into non-human concepts of self would be our scientific understanding of nervous systems. Any creature with a nervous system demonstrates an understanding of a singular point orientation in relation to the environment. All creatures with a measurable nervous system behave in ways that protect such vantage point. At the very least, we can posit that this demonstrates an exceptionally primitive sense of self through self-preservation.

So, with a pragmatic lens in mind, the observable universe allows us to identify consciousness as "awareness of your surroundings, the ability to create, **demonstration** of awareness by the mind of itself, and an imaginable existential experience." Note that what we have done is identify consciousness as more of an act rather than a substance. This delineation is not because consciousness is only a verb but because consciousness is a fluid causal essence. Its manifestation is analogous to being water in a glass, except in the element of consciousness being able to play an emanating causal role in the process of creation. It is similar in that its substance is completely malleable to the form of life that encompasses it. In different species, the substance is the same, but the magnitude, degree, or level of consciousness becomes more sophisticated within more evolved mental capacities like the dolphin, ape, or human being. It has taken us quite some time to figure out that the **investigative dive into biological studies is synonymous with gazing into prior forms of consciousness**.

Conscious experience seems distinct from the human vantage point, however; it is comparable to any other biological element in that it plays a fundamental functional evolutionary role in the formation and sustainment of life. As we observe from the most fundamental forms of agency toward the phenomenon of human conscious experience, a specific pattern reveals itself.

In the great evolutionary lineage of species, agency and the emanation of awareness quantitatively increases through the expansion of each species through the benchmark of possible outcomes. From the hydra to the flatworm, the Castoridae to the beaver, and the ape to the human, the conscious capacity through the lineage of evolution is quite apparent. If we isolate a troop of apes, they will formulate their own social structure around things of value; given that they are so close to our species' genetic lineage, it is easy to see the analogous behavior between the two species. However, apes will not show the capacity to build social structures around something as complex or capable as the modern human economy. Through the elaborate and clever mechanisms that make human lifestyle appear different, you could mistakenly understand consciousness as an exclusively human quality. The idea that we have transcended the substantial form of natural existential challenges is no more than a mere illusion. We must realize we are in an evolutionary stage of conscious transcendence, not a complete realization of conscious transcendence. The point is that even though we are more consciously capable relative to the species that have come before us, we are not distinctly unique enough to say we are the only species that possesses consciousness, nor will we be the 'most' conscious being for all of time.

Through the lens of evolutionary biology, philosophy is not rendered mundane, philosophy is reinvigorated. With the theory of evolution, the study of phenomenology is equipped with greater material to synthesize. The theory of evolution allows us to philosophically posit that evolution is the act of conscious expansion of itself, by itself. The fluid expansion of consciousness, of itself, by itself, raises a familiar tone spoken by Hegel, who I have already alluded to in this book. Hegel's *Phenomenology of Spirit* is a uniquely subterranean text; its rendition of phenomenology includes a wise synthesis of contextual and orientationally cognizant information. As I introduced earlier in the text, the process of the natural world takes the procedural form of his proposed dialectic. By understanding Hegel's message with the theory of evolution in mind, the phenomenological dialectic materializes further. The dialectic of becoming is intimately embedded within the mechanical process of evolution, but the dialectic is not simply a process, it is an omnipresent transcendence. The point of the dialectic is to point out that there is always a becoming of the self by the self. When we apply the dialectic to the human, we can observe a transcendent nature of every conscious action. In the simple act of a human creating the water container, we didn't just become a bipedal animal with a water container. **We became a bipedal animal synthesizing its being with the water container**, and this measurably led to human transcendence into becoming a more mobile form with the phenotypic formation of arches in the feet. **This same principle applies today; we are the bipedal animal synthesizing and transcending with cyborgian technological creations of the conscious mind.**

The human being is actively transcending with greater efficiency because of the creation of additional lines of inheritance and an expanding environment. These are the critical actors of our existential context to understand. Since the beginning of culture, the collective conscious mind has manufactured a collectively remembered article of transmitted information. Cultural transmission comes in addition to the collectively formed genetic memory we also inherit. In essence, even the nature of human nature was founded in the workings of a conscious subjective mind. In these words, I am emanating conscious awareness to the conscious mind of the role that the conscious mind plays. This is the self-realization of consciousness to itself on a much larger scale than the typical conscious phenomenon, which is the self-realization of the self.

Furthermore, with the idea of transcendence in mind, the temporal scope of consideration causes us to expand our gaze toward what this means for the future. In the context of evolution, consciousness is growing congruently with the expansion of sense and mobility in relation to the holistic environment. With the expansion of these qualities, variability, potentiality, and ascendency all expand. All three of these qualities indicate an underlying abstract concept actively being expanded and amplified. I posit that this abstraction of the theory of evolution indicates a teleological expansion toward the realization of "freedom."

Historically unanswerable questions become answerable when applying the new evolutionary lens toward traditional philosophical considerations. The philosophical debate over the status of free will in humanity is an example of such traditional considerations that become answerable with this new evolutionary lens. Philosophy has considered the debate over human capacity to free will to be primarily unanswerable, but the contextual understanding of evolution provides an understanding into the truth of the matter. To summarize briefly: The debate over this matter has been between determinism and free will. Determinists claim that all action is determined by environmental factors apart from the will or conscious intervention and that freedom of will can be nothing more than an illusion. The free will position naively states that the human will can perform actions and make decisions independent of external worldly influences. The best way to understand this debate is by looking at the world in terms of causality. Was my will to choose the determinant factor in deciding not to subscribe to a particular religion? Favorite foods? Or even my political beliefs? Or is there some genetic influence or subliminal messaging in my environment that determined my decisions?

Yet again, free will has been unanswerable because of an insufficient breadth of temporal scope considered in the debate. If the conscious awareness of the mind is itself a determinant factor in the environment, then determinism is impossible in the context of a human. If each constituent part of the environment that beholds the conscious mind mechanistically shapes the conscious mind, then free will is impossible for the human. However, through the history of evolution, we can observe the expansion of sensuous, mobile, and cognitive capacities influencing the advancement of each other. With the advancement in mobility, the conscious environment expands, with advancement in sense, the conscious environment expands, and with advancement in knowledge, the conscious environment expands. These elements of a given being determine the degree to which that being is consciously free. As these elements of evolution grow, consciousness capacity consequently grows. Thus demonstrating, as these properties expand in species through eons of evolution, the freedom of the conscious will concurrently expand. As species have evolved, the environment actively loses deterministic capacity at each stage of the evolutionary timeline. So then, the conscious expansion of the universe has a defined direction. Evolution is not aimless; it entails a foreseeable purpose. **The evolution of species is actively moving toward the realization of a species that realizes complete freedom of will. Thus, evolution is not simply a biological process but a teleological phenomenon of conscious expansion.**

To the nihilist that believes there is no point in your actions, you have failed to understand the world. The purpose of your being originates from the stars, and you are the realization of the universe realizing itself and its purpose. Through understanding this purpose, you have the power to not only be but to be actively transcendent and aware of your transcendence in the becoming of the universe. **You are a profound, essential role in unfolding this universal undertaking since the known formation of life.** When you observe your being, observe your being within the natural world as the natural world. The human mind's habit of formulating orientational barriers is entirely misguided. Our categorical delineations within language, especially in understanding the word "nature," has disoriented us with the truth of our essence. When you observe your context, feel the unfolding of action around you. There is an unstoppable dissipation of action toward an end. We now have a better understanding of this end; respect the work that has been done before you and is being done before you. The responsibility is on us to recapture the purpose of nature and be the purpose of nature. Because this still needs to be done, we do not have a realized form or conception of free will. Instead, we have been narrow-mindedly debating the false dichotomy of determinism and complete free will. However, with a deeper understanding of consciousness in evolution, a true depiction of an empirically derived God begins to take shape.

DEFINING GOD

"Never lose an opportunity to see anything beautiful, for beauty is in God's handwriting. ... Standing on the bare ground,—my head bathed by the blithe air, and uplifted into infinite space,—all mean egotism vanishes. I become a transparent eye-ball; I am nothing; I see all; the currents of the Universal Being circulate through me; I am part or particle of God."
–Ralph Waldo Emerson

Deeply understanding the biologically universal process of evolution is central to understanding the existential context of the human being in pursuit of deeper meaning. Evolution is frequently posited as an adequate replacement for historical conceptions of God. This substitution is not a completely misguided understanding of reality, but there are several mistakes associated with this claim. The primary mistake is in isolating the word "God" to refer solely to the origin/creator of humanity. The other mistake is in applying the standardized conception of what the theory of evolution represents. The standardized conception of evolution is still philosophically primitive as it has only been exercised as a mechanism that biologically explicates our origins but leaves a void in application toward deeper meaning. If we are to claim that evolution sufficiently replaces God, then we must understand evolution more thoroughly, with consideration for what 'God' really means. Evolution is sufficient alone in establishing a source of creation, but "God" is an idea that encapsulates much more than that.

The term "God" has been an essential cultural reference point since recorded language began. The human understanding of "God" has been used to explain many essential elements that encapsulate the existential essence of humanity. For example, the word "God" in modern American culture traditionally insights a conception of a personified old, white-bearded figure who begets and oversees the happenings of the world. This cultural understanding exemplifies a monotheistic tradition, seen in Judaism, Christianity, and Islamic teachings. This depiction of God is typically defined as "the **creator** and **ruler** of the universe and source of all **moral authority**; the supreme being." (Oxford Languages). In some religious beliefs, God is defined as "a superhuman being or spirit worshiped as having **power over nature or human fortunes**; a deity." (Oxford Languages).

The perceived meaning of the word varies slightly depending on the sampled human culture we observe. If you are to refer to "God" in ancient Greek, ancient Egyptian, or ancient Nordic culture, you could be referring to a deity among a multitude of deities. In these polytheistic traditions, the word "God" still adheres to the definitions cited above, the only difference being that you must name which deity you are referencing, or the reference will be futile. "God" has also been standardized to personify the fate of a given person's life. For example, the phrases: "God willing" and "God relenting." In these phrases, it is implied that God is the ultimate determinant of whatever elements unfold. May this usage be the most glaring evidence of our chosen ignorance regarding our orientation with our capacities. Another standardized usage of the word is embodied by appropriating an item as "God" if the item is demonstrated to be of absolute importance. This usage is exemplified by the sentence, "money is her God." These standardized usages perpetuate the traditional definitions referenced above, thus demonstrating that the word's meaning has been preserved through many different generations of different cultures.

As we look at the consistent elements that the term "God" represents cross-culturally, we see this word represents more than just the origin of our creation. By extracting the consistent attributes from each of the definitions and uses above, we can extract that "God" represents: **Creator, moral authority/purpose, and power over nature and human fate**. The elements of each chapter in this book put forward empirical understandings that have built a new way of understanding the biological process of evolution. Through a deeper understanding of evolution, humankind can now empirically determine what the word "God" represents.

The mistake in the standardized conception of evolution is that it is typically isolated as an autonomous mechanistic procedure for nothing more than environmentally adaptive gene mutations in "real time." How evolution has been presented in the previous chapter presents its deeper *nature*. Namely, all evolved features of nature have evolved because *something* became **aware** of a deficiency or environmental need which led to an awareness of newfound possibilities that increase survivability. Through this process, evolution has observably produced an exponential increase in awareness through the lineage of species that contributes to the formation of the next. Therefore, consciousness is the primary catalyst of creation that manifests its own expansion through the evolutionary process. To be clear, consciousness alone is not **the** creator, but the driver, the spirit, the conduit underneath the physical process of creation. Though consciousness is not the creator, I firmly posit that consciousness is the most necessary component of creation. The role of consciousness within the process of evolution is the most integral element responsible for the origins of creation. Therefore, consciousness as a conduit in the evolutionary process explains our origins more thoroughly. Here we have a more nuanced understanding of how evolution, a teleological phenomenon, practically materializes as the explanation of origins/creation in a "Godlier" sense. By understanding evolution with a deeper consideration of consciousness, consciousness satisfies this essential component within the definition of "God" as the **creator**.

As species evolve, the increase in consciousness is the consistent result of the observable evolutionary progression. The realization of consciousness is emerging from nature through a transcendent synthesis of itself, by itself. From this process, a teleological advancement of conscious freedom will likely result from evolution as it has for its entire observable history. The purpose of extracting the role of consciousness is not just for the sake of understanding evolution from a different angle. It is for the sake of understanding an underlying **purpose** of our being that has been sitting right in front of us.

Identifying evolution as a teleological phenomenon of realizing conscious freedom brings forth another vital implication. The continued measurable increase in conscious awareness through the lineage of species submits that we are **purposed** to continue this lineage, especially now that we can claim awareness of this phenomenon. Due to our self-distancing from nature through the continued use of the poorly defined definition of the word "nature," this element of consciousness we have understood to be an integral constituent part that defines humanity is an intimate part of all living beings. This concurrence entails a more important element of nature to understand. To our knowledge, we possess the most evolved mental and technical capacity, which measurably gives us the most freedom of will relative to other living creatures. Therefore, we must understand that an increase in freedom entails an increase in responsibility.

Evidence of gene-culture evolution shows that our conscious action determines creation more than any other species. Our consciousness has become the most integral part of the dialectical evolutionary mechanism. As we know from the example of the human invention of the water jug leading to the development of arches, our involvement in this process is incredibly consequential toward future lines of inheritance. No matter what action we take, there will always be genetic, environmental, and cultural consequences for our next line of inheritance. Since we possess this magnitude of conscious freedom and ascendent capacity, we are coming into a newfound realization of **authority**.

Given this, we have the power to determine our purpose: to ensure our actions change the coming lines of inheritance for the best. Here, we can revisit the qualitative metric as our guide to determine what "best" means. The qualitative metric establishes that orienting ourselves with meaning requires understanding what provides the most utility, what has the highest capacity to parallel the variable nature of the world, and what expands our proximal senses, thus **expanding our awareness**.

Let our **purpose,** and thusly **moral authority,** be determined by the values of discipline in the qualitative metric, an emanation of awareness, and the preservation of that which has become aware and upholding that which is aware. Here, the Godly qualities of **purpose and moral authority** as a consequence of simply existing within the conscious medium of humanity reveal themselves through the more profound understanding of biological evolution.

It is relevant to mention that human-made biological technology allows us to manipulate, alter, and create genetic sequencing over our DNA in real time. For example, DNA sequencing gives us the power to alter and create biological life in a single stroke, a power historically reserved for many cultural and religious definitions of God. However, this doesn't change the fact that when we observe the history of evolution, consciousness' ascendency over genetic mutations has always been the reality; it was never as apparent to our temporally perceivable field as it is now. In this, humanity is not distinctly God-like. We just possess a higher magnitude of ascendent power relative to other forms of consciousness.

Again, this newfound ascendent freedom requires us to familiarize ourselves with profound responsibility congruent with this profound power. The moral priority of remaining disciplined to the qualitative metric, preserving consciousness, and preserving the elements which biologically sustain it, is the basis of what we are responsible for upholding. Because consciousness has the most consequential impact on the teleological phenomenon of evolution, we must make its preservation a moral priority. And as we have come to understand a deeper understanding of *nature*, we have come to understand that *nature* and consciousness are not distinct. Therefore, we must preserve the biological elements in what we traditionally understood to be "nature" because they are the constituent parts of the indistinct whole that upholds the conscious phenomenon. Remember the earlier example of the tree; when isolating the metric of "sustaining life," our lungs, surrounding trees, and plants are integral in upholding this quality of sustaining our life. We must be intentional and authoritative in reciprocating this favor to all environmental components that sustain our consciousness. It is not only the duty to continue the lineage of conscious expansion, but to preserve the existent life that has already transubstantiated into this world.

As we know, genetic evolution is reactionary to the environment, yet we have completely manufactured entirely new environments the history of species has never seen before. Almost every element of the conscious human environment was functionally created through human conscious intervention. Consider language, technology, shelter, knowledge, belief, and culture; which of these elements did not originate from the conscious subjective mind? Where else would interpretations of religious doctrines and cultural dogmas that describe purpose come from, if not the subjective conscious mind? Our culturally manipulated environments have been the determinant of genetic adaptations we are fated to possess. Therefore, the conscious mind has entirely manufactured what our genes will respond to. This entails that humanity and humanly manufactured environments determine our fate and the ensuing species' **fate**.

The human mind's material participation in the evolutionary process explicitly satisfies the cross-cultural qualities of God, namely: **Creator, moral authority/purpose, and power over nature and human fate**. Empirically, God is not a deity figure but an act of transcendence that takes the form of the essential dialectical mechanism between consciousness, genes, and the environment. This dialectic is now overwhelmingly controlled by our consciousness, thus bestowing the understanding of God to the materialization of what our consciousness decrees. Our human consciousness is now solely responsible for determining the conditions that create the next being in the evolutionary chain, solely responsible for declaring moral authority implied in this reality, solely responsible for determining our purpose, and solely responsible for the **fate** of our own and succeeding species. These claims reflect the empirically derived nature of the context we occupy as human beings and the cross-cultural definition of "God." Given this responsibility, we must revisit the importance of the empirical discipline we have built through the qualitative metric. Our consciousness possesses remarkable responsibility, but we still have environmental resources that can functionally guide our movement toward realizing freedom, regardless of how dizzy our freedom makes us.

So then, within the boundaries of our collective knowledge, this is the best formulation of who God is. Observations I have communicated throughout this book may have breached the taught boundaries of many religious understandings, but it is not meant to be exclusory to those who subscribe to such religions. It is important to consider that any religious doctrine could not have been perfectly synthesized or communicated through time. The world changes through time, so how could a solidified religious doctrine find truth if the world is consistently changing? We must build a doctrine that remains congruent with the everchanging world and adapt to observable truth. The observable nature of our standing in nature, as nature, moving nature, is our truth. It is our duty to derive meaning and purpose from this observable position in reality. Observable truth renders nihilistic conceptions of the world to be completely arbitrary as one considers the empirically deducible power, responsibility, and purpose of each human being. We are tasked with bestowing increased freedom to the next generation of consciousness.

In the current phase of nature, we stand alone as the cause of all future inherited lines of reality. Look no further than your own gaze to understand God. The utmost form of God is you.

REFERENCES

Baggini, J. (2002). Bertrand Russell: The Problems of Philosophy (1912). In Philosophy: Key Texts (pp. 85-114). London: Palgrave Macmillan UK.

Descartes, R. (2002). Meditations on First Philosophy, Meditation II. *The European Philosophers from Descartes to Nietzsche, New York: The Modern Library.*

Hegel, G. W. F. (2018). *Hegel: The phenomenology of spirit.* Oxford University Press.

Musser, J. M., Schippers, K. J., Nickel, M., Mizzon, G., Kohn, A. B., Pape, C., ... & Arendt, D. (2021). Profiling cellular diversity in sponges informs animal cell type and nervous system evolution. *Science, 374*(6568), 717-723.

Nagel, T. (1974). What is it like to be a bat?. The philosophical review, 83(4), 435-450.

Nietzsche, F. (1998). *On the genealogy of morality.* Hackett Publishing.

ACKNOWLEDGMENTS

I would be remiss if I didn't express my gratitude toward those who helped me write this book. Firstly, I am incredibly grateful for my spouse, Dr. Jayne Simpson. Thank you for challenging me and my conceptions of this world every day. Thank you for your endless teamwork and support. I couldn't have finished this book without you. Thank you to Dr. Myles Mason for editing this book; you helped me clarify my words and thoughts immensely. Thank you to my mother for balancing many responsibilities while homeschooling me in my developing years. Thank you to my father, Craig Tucker, for modeling work ethic, accountability, and ethics. Thank you to my friends Michael and Derek for offering so many valuable exchanges of thoughts. Thank you to my siblings, Breanna, Hanna, Luke, and Anna, for magnificently impacting my life.

ABOUT THE AUTHOR

Casey Tucker holds a B.A. in Philosophy from Stanford University. However, Tucker's investigations into the natural mysteries of this world neither started nor ended at Stanford. From homeschooling in isolation to Tucker's postgraduate freelance studies, he has a passion for engaging with the edges of what humankind claims to 'know.' This book emerged from the synthesis of many different philosophical texts, emerging scientific theories, and the human propensity to wonder.